Divided Spaces, Contested Pasts

The Gallipoli peninsula in Turkey was the site of one of the most tragic and memorable battles of the twentieth century, with the Turks fighting the ANZAC (Australian New Zealand Army Corps) and soldiers from fifteen other countries.

This book explores the history of its landscape, its people, and its heritage, from the day that the defeated Allied troops of World War One evacuated the peninsula in January 1916 to the present. It examines how the wartime heritage of this region, both tangible and intangible, is currently being redefined by the Turkish state to bring more of a faith-based approach to the secularist narratives about the origins of the country. It provides a timely and fascinating look at what has happened in the last century to a landscape that was devastated and emptied of its inhabitants at the end of World War One, how it recovered, and why this geography continues to be a site of contested heritage.

This book will be a key text for scholars of cultural and historical geography, Ottoman and World War One archaeology, architectural history, commemorative and conflict studies, European military history, critical heritage studies, politics, and international relations.

Lucienne Thys-Şenocak is a professor of cultural heritage management, museum studies, and architectural history in the Department of Archaeology and the History of Art at Koç University in Istanbul, where she has been teaching for 25 years. She has worked on several heritage projects in the Gallipoli region since 1997 and was the co-director of the team which began the documentation of archaeological and conservation research at the Ottoman fortress at Seddülbahir, located on the Gallipoli peninsula at Cape Helles. In addition to the cultural heritage of the Gallipoli peninsula, Thys-Şenocak has published on the architectural patronage of Ottoman royal women, and the heritage of viticulture in Anatolia. She currently advises the Çanakkale Wars and Gallipoli Historic Region Directorate for the restoration and museum project at Seddülbahir, which is expected to be completed and open to the public in 2019. She is a member of the International Council of Monuments and Sites (ICOMOS) in Turkey.

Divided Spaces, Contested Pasts

The Heritage of the Gallipoli Peninsula

Lucienne Thys-Şenocak

Routledge
Taylor & Francis Group
LONDON AND NEW YORK

First published 2019
by Routledge
2 Park Square, Milton Park, Abingdon, Oxon OX14 4RN

and by Routledge
52 Vanderbilt Avenue, New York, NY 10017, USA

First issued in paperback 2020

Routledge is an imprint of the Taylor & Francis Group, an informa business

British Library Cataloguing-in-Publication Data
A catalogue record for this book is available from the British Library

Library of Congress Cataloging-in-Publication Data
A catalog record for this book has been requested

ISBN 13: 978-0-367-58725-3 (pbk)
ISBN 13: 978-1-4724-1446-5 (hbk)

Typeset in Times New Roman
by Florence Production Ltd, Stoodleigh, Devon, UK

For my family

Contents

Illustrations

Figures

Tables

Acknowledgements

I began writing this book in 2012, but in many ways, it represents twenty years of engagement with the Gallipoli Peninsula. Since 1997, the year I started researching the Ottoman fortress of Seddülbahir in the Cape Helles section of the peninsula, I have learned a great deal about this landscape and its heritage from many who live in the Gallipoli region and/or work on different aspects of its past and present. No one I have met or worked with is neutral about Gallipoli. It is a place which has relatively unexplored archaeological layers, and an intensively investigated identity as a battlefield of World War One. As a site of wartime tragedy, and epic military failures and successes, the Gallipoli Peninsula continues to inspire those who encounter it to remember and honor the many lives that were lost there. Particularly since the end of World War One, commemorative actions have so often been expressed as architectural projects. As an architectural historian involved in the preservation of tangible heritage, I found that the buildings—Ottoman and those from more recent times—were what first captivated my interest in Gallipoli. Now, however, these are cause for concern. The past century has seen so much of the landscape of the Gallipoli Peninsula marked by architectural projects, commemorative and otherwise. However well-intentioned, these projects are rapidly transforming the landscape of this region, often before what is lost can be documented. Landscapes, like all elements of heritage, tangible and intangible, are always changing and represent a complex web of interactions between humans and nature. The challenge for all who are passionate about Gallipoli, then, is to steward it carefully through the next century. I thank here everyone with whom I work and share the goals of mindful stewardship of this unique cultural landscape.

That responsibility is now officially in the hands of the current administrators of the Gallipoli Peninsula: the Çanakkale Wars and Gallipoli Historic Region Directorate, or ÇATAB. I thank the past and present directors, Mehmet Gürkan and İsmail Kaşdemir, respectively, for inviting me to participate in many of the strategic planning workshops for the Gallipoli Peninsula from 2014–2017. Particular thanks go to the dynamic team from ÇATAB who are restoring the fortress of Seddülbahir, and to my colleagues from Istanbul Technical University with whom I have worked for so many years at that site: Rahmi N. Çelik, Gülsün Tanyeli, and Arzu Özsavaşcı. I would also like to express my gratitude to many of the faculty and students at Koç University who have listened to my ideas about

Gallipoli take shape over the past two decades, and have shared theirs with me. Carolyn Aslan kindly provided her important insights and experience about the archaeology of the Gallipoli Peninsula; Gizem Dörter helped compile the chart of archaeologists working on the Gallipoli peninsula; Bengi Atun assisted by organizing notes from ÇATAB workshops. Elifgül Doğan made several of the maps; Furkan İnan and Edip Gölbaşı helped with photography and archival research.

I have used many of my own images and engravings to illustrate this book, but have also benefitted immensely from many archival collections in Turkey and abroad. In particular, I would like to thank the Commonwealth War Graves Commission Archives in Maidenhead, and Roy Hemington for assisting me with my research there. I am also grateful to Hugh Miller, one of the US National Park planners who worked in Turkey in the late 1960s, for sharing his experiences with me about that time. Ertuğrul Ceylan, a lawyer on the planning team for the Çanakkale Martyrs Memorial building campaign in the 1950s, kindly agreed to be interviewed about that project.

I could never have completed this book without the editorial expertise of Victoria Barnes, the entire team at Routledge, and the unflagging support of my husband, daughters, sisters, mother, and so many friends and family members. You encouraged me to keep researching and writing, and to finish what I had started, "bird-by-bird," and I am so grateful to each and every one of you for that.

Abbreviations, information on archival sources, websites, and Turkish words

AGAD Anzac Gallipoli Archaeology Database

AKP Adalet ve Kalkınma Partisi (Justice and Development Party)

ANZAC Australian and New Zealand Army Corps

AWM Australian War Memorial

BCA T. C. Başbakanlık Cumhuriyet Arşivi (Turkish Republic, Prime Ministry Republican Archives, Ankara)

BOA T. C. Başbakanlık Osmanlı Arşivi (Turkish Republic, Prime Ministry Ottoman Archives, Istanbul) and in the BOA the following: DH.UM VM (Dahiliye Nezâreti Umûm Vilâyât Müdüriyeti), MD (Mühimme Defterler), MMD (Maliyeden Müdevver Defterler)

CWGC Commonwealth War Graves Commission

ÇATAB Çanakkale Savaşları ve Gelibolu Tarihi Alan Başkanlığı (Çanakkale Wars and Gallipoli Historic Region Directorate)

GIS Geographical Information System

GPS Global Positioning System

GPR Ground Penetrating Radar

GRU Graves Registration Unit (became the Graves Registration Commission in 1915)

IWM Imperial War Museum

IWGC Imperial War Graves Commission

ITU Istanbul Technical University

JHAS Joint Historical and Archaeological Survey

METU Middle East Technical University (abbreviated in Turkish as ODTÜ)

OPET Name of a private Turkish petroleum company

TNA The National Archives, UK; in the TNA the CAB (Cabinet) FO (Foreign Office) and WO (War Office)

TSMA Topkapı Sarayı Müzesi Arşivi (Topkapı Palace Museum Archives)

TTKD Türkiye Tabiatını Koruma Derneği (Foundation for Turkish Nature Conservation)

TMMOB Türk Mühendis ve Mimar Odaları Birliği (The Chamber of Turkish Engineers and Architects)

UDGP	Uzun Devreli Gelişme Planı (Long-term Development Plan)
UNESCO	United Nations Educational, Scientific and Cultural Organization
USAID	United States Agency for International Development
USNPS	United States National Park Service

Information on archival sources

Several, but not all, of the archival documents from the CWGC were digitized and assigned new reference numbers during the time I was researching and writing this book. Whenever possible the original archival reference (Alt Ref No.) is given first, followed by the Unique ID number of the recently digitized document, and separated by a semicolon. As an example, the archival documents about French Graves in Egypt and Gallipoli can be found at WG 1188/6; CWGC/1/1/8/15. Many of the archival documents cited from the Australian War Memorial Archives are from the AWM 38 series "Official History, 1914–18 War: Records of C. E. W. Bean, Official Historian" and are cited as AWM38 3DRL 8042-51, unless otherwise indicated.

Websites

Websites for digital versions of cited sources, or links to websites where more information is available, are occasionally provided in the endnotes. Frequently cited websites also appear in the bibliography. All websites were accessible as of 15 June 2018.

Turkish words

Modern Turkish orthography is used for all words that appear in Turkish, with the exception of a few words that are commonly used in English, such as pasha, haj, or Quran. "Istanbul" is spelled with an undotted capital I in the main text, but a dotted capital İ is used in the endnotes and bibliography if this refers to a place of publication.

Figure 0.1 Satellite map of the Gallipoli peninsula and surrounding area, 21 March 2006.

Credit: NASA and Google Earth

Figure 0.2 Detail of engraving of Gallipoli Peninsula, *Carte de l'Hellespont*, Choiseul-Gouffier, 1789.

Credit: Archive of L. Thys-Şenocak

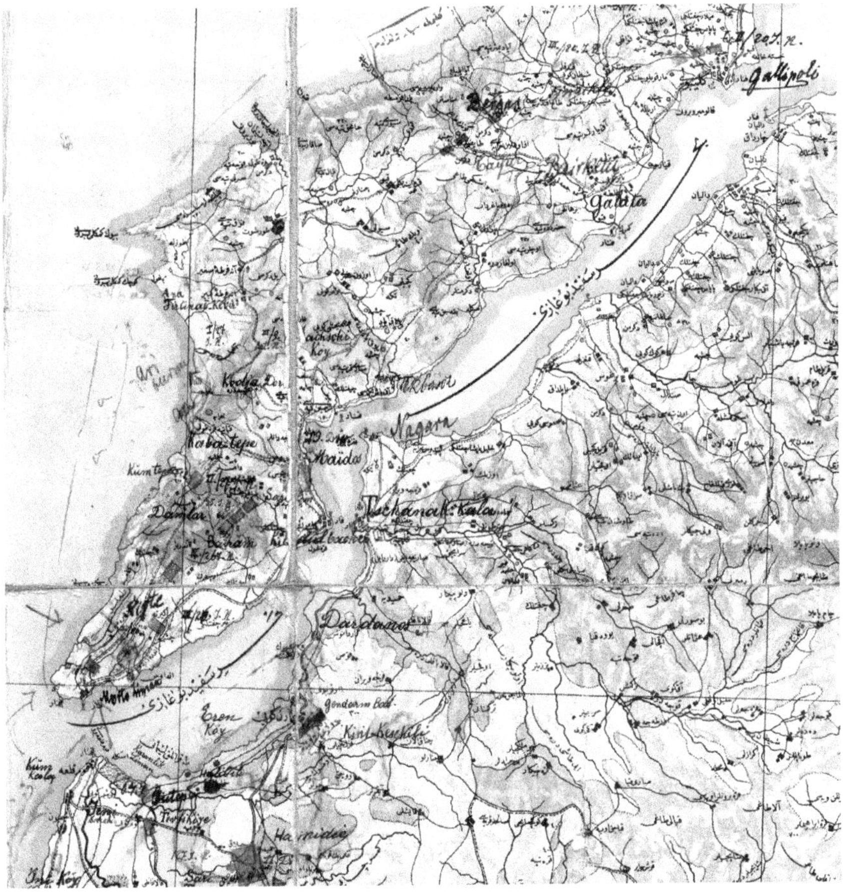

Figure 0.3 Gallipoli Peninsula. Detail of Ottoman map showing 25 April 1915 landing points, (signed "Liman von Sanders V Armée").

Credit: AWM PO7906.149

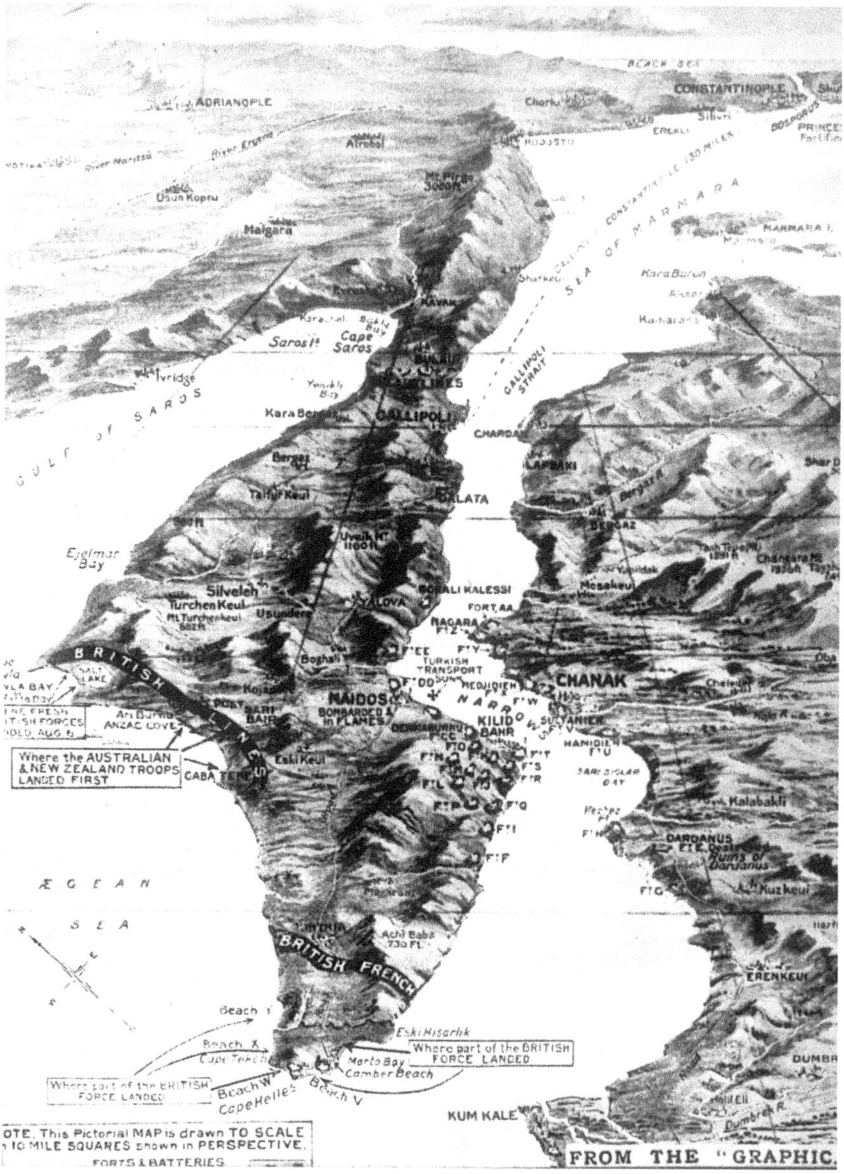

Figure 0.4 Front page map of the Gallipoli peninsula during World War One showing British and French positions from the *Graphic*.

Credit: Imperial War Museum, London Q70855

1 Introduction

The Gallipoli peninsula is a spit of land, 33,500 hectares and 80 kilometers long, which is located on the North Anatolian Fault zone and forms part of the Alpine Pontide range. It delineates the northern border of the Dardanelles, the waterway that connects the Marmara Sea to the Aegean, while separating the two continents of Europe and Asia along the fracture between the European and Anatolian tectonic plates. The peninsula takes its name from Ancient Kallipolis, "the Beautiful City," which is now the Turkish town of Gallipoli located at the far northern entrance to the peninsula.[1] Across the Dardanelles from the Gallipoli peninsula is the Troad, the lands containing Priam's legendary city of Troy, now an archaeological park and World Heritage site.[2]

Today in Turkey the peninsula is referred to as the Gelibolu Yarımadası—literally, the "half island" of Gallipoli. The strait, which at its narrowest point is twelve hundred meters, is known as the Çanakkale Boğazı, or the "throat" of Çanakkale, and is the location of the largest city on the Asian shores of the Dardanelles; in 2017 it was home to half a million residents. The construction of the Çanakkale 1915 bridge, which will connect the European and Asian shores of the Dardanelles, began on 18 March 2017, the anniversary of a symbolic Ottoman naval victory in the Çanakkale Wars. The world's longest suspension bridge is expected to be completed by 18 March 2022 and is intended by the government as a major stimulus for the development of tourism and industry in the region.

In the sixth century BCE the Gallipoli peninsula was referred to as the Chersonese, and the waterway that separates the Asian and European continents was known as Hellespontos, or the Sea of Helle, after the Classical mythological figure of Helle, a consort of Poseidon. Helle and her brother Phrixus were assisted by their mother to avoid the wrath of an evil stepmother by escaping on the back of a flying golden ram; but Helle lost her grip, fell to her death and drowned in the waters of the straits (Figure 1.1). While we lack sufficient archaeological data from the region to provide a comprehensive chronology, the surveys and archaeological excavations that have been undertaken on the peninsula in the past century indicate that humans have lived there since the Prehistoric era.[3] The Antique sites on the shore of the straits facing Asia, such as Elaious, are mentioned in the writings of many Classical and Roman-era chroniclers and geographers. Excavated by the French during World War One, and again between 1920 and 1923, the necropolis of Elaious revealed an abundance of artifacts and sarcophagi.[4]

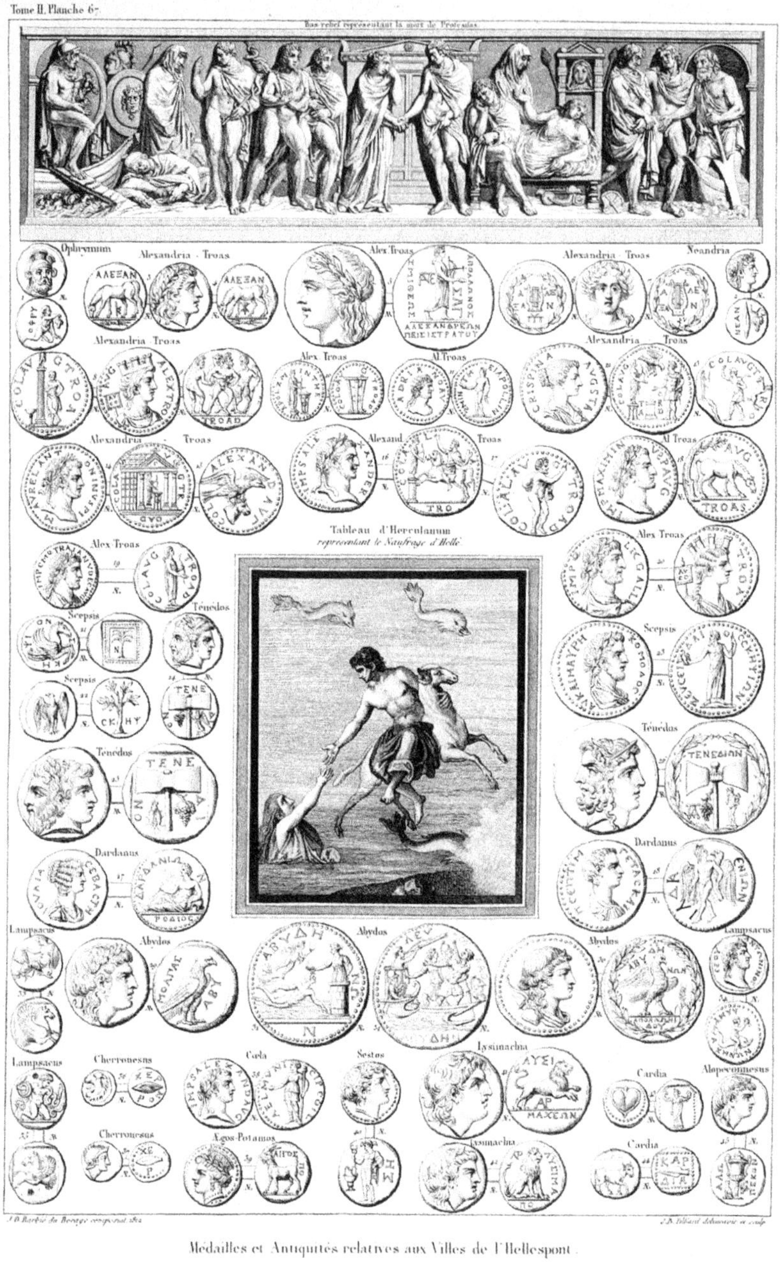

Figure 1.1 Image in the center shows the fall of Helle into the sea, surrounded by the coins of the ancient cities around the Hellespont. From "Médailles et Antiquités relatives aux Villes de l'Hellespont" in Choiseul-Gouffier's *Voyage pittoresque de la Grèce*, Paris, J.-J. Blaise, 1822.

Credit: Aikaterini Laskaridis Foundation

Famous crossings of the straits were undertaken by the armies of Persian warrior kings Darius I in 514 BCE and Xerxes in 480 BCE, the latter with boats strapped together to create a pontoon bridge for his invading troops. In 334 BCE, Alexander the Great, while conquering Thrace and then the Gallipoli peninsula, is alleged to have made a pilgrimage to a remote site at the end of the peninsula to pay homage to the mythic Greek hero Protosilaus, among the first to fall in the Trojan War.[5] Later Byzantine and Ottoman rulers recognized the strategic value of the peninsula and erected fortresses and customs houses along its shores to keep pirates and enemies, such as the Venetians and later the Russians, at a safe distance from the capital of Constantinople; and to collect revenue from the lucrative trade that passed from the Black Sea to the Mediterranean and back.[6]

Due to its strategic location bordering the Dardanelles, the Gallipoli peninsula has been the site of many conflicts in its long past. However, it is best known as the location of one of the major battles fought in World War One by the Allied forces—comprised of the French and Commonwealth troops—against the Ottomans and their German allies. *Divided Spaces, Contested Pasts: The Heritage of the Gallipoli Peninsula* examines the heritage of the post-WWI landscape, its archaeological and commemorative projects, and the political agendas which have shaped the past and the present of the Gallipoli peninsula since 9 January 1916, the day that marked the evacuation from the peninsula of the last of the defeated Allied forces.

The Gallipoli peninsula: a brief history since World War One

Conflicts between the Ottomans and the Allied forces during World War One began in November of 1914 when the coastal fortresses of Seddülbahir and Kumkale, part of the defense system at the Aegean entrance to the Dardanelles, were bombarded. The actual land invasion that has come to symbolize the beginning of the Gallipoli campaign in World War One (known as the Çanakkale Wars in Turkey), and the day that is commemorated by Australians and New Zealanders as ANZAC Day, was 25 April 1915.[7] The main reason for the invasion of Allied forces in this southern European theater of war was that the offensive against Germany on the European Western Front had ground to a standstill early in 1915. The intention of the Allied forces was to open a new front in the east and establish a safe supply line between their ally Russia and the Mediterranean Sea. Russian access to this warm water port was crucial for the Allied troops as the Tsar's munitions and shipments of grain were essential to the success and sustenance of their European armies. Both Lord Kitchener, British Secretary of State for War, and Winston Churchill, First Lord of the Admiralty, supported a plan to attack the Gallipoli peninsula. This strategy, however, required that the Allied navy first capture the straits of the Dardanelles and then move north to the Bosphorus to take the capital of the Ottoman Empire, known then as Konstantiniyye, or Constantinople.

The Allied forces did win World War One, but the plan to take the Gallipoli peninsula from the Ottomans failed and has been recorded in history as a disaster of epic proportions in terms of its poorly planned military strategy, implementation, and resulting casualties. After nine months of fierce fighting on the peninsula between the Allied and Ottoman troops, the battles ended, the Ottomans were victorious, and the Allied troops quietly evacuated. The total number of casualties in the Gallipoli battles was close to half a million. The exact number of dead is still debated on all sides, but estimates are that at least 22,000 British and Irish, 7,500 Australians, 2,700 New Zealanders, and 10,000 French died on the Allied side. The majority of the Allied troops had been conscripted from England and France but there were also soldiers from Canada, Ireland, Scotland, Wales, India, Malta, Tunisia, Algeria, Gambia, and Senegal who were part of the Allied forces. On the opposing side, up to 130,000 Ottoman soldiers are thought to have perished during the years of combat.[8] Among the Ottoman troops there were also soldiers of diverse ancestry, including Arabs, Greeks, Armenians, Kurds, and Jews.

The Greek and Turkish civilian populations living in the villages on the Gallipoli peninsula either fled or were evacuated just prior to the major battles that began there in 1915. The few Greeks who returned to their homes in the villages on the peninsula when World War One ended on 11 November 1918 did so for a short time only; most emigrated permanently after the major Greek defeat in August 1922 during the Turkish War of Independence.[9] After the signing of the Treaty of Lausanne on 24 July 1923, problems related to the exchange of peoples and properties between Greece and Turkey, known as the 1923–24 Population Exchange (*Türkiye-Yunanistan Nüfus Mübadelesi*), were extensive and the archives are full of court suits on both sides from plaintiffs, some of whose fields and homes were expropriated by the Allied or Turkish governments to use as land for commemorative monuments and cemeteries.

Identification, exhumation, reburials, and commemoration of those who had died on the Gallipoli peninsula during World War One occurred during the war, but both the Allied countries and the Ottomans began a more organized and official process of dealing with their dead in 1918, after World War One had ended. By 1915 the British had formed the Graves Registration Commission as part of the Red Cross; the former organization would evolve in 1917 into the Imperial War Graves Commission (IWGC) and subsequently into the Commonwealth War Graves Commission (CWGC), which by 1926 had concluded a formidable project on the Gallipoli peninsula resulting in 31 cemeteries and 6 commemorative monuments complete with extensive landscaping. To date, close to 23,000 graves have been individually marked in the many CWGC cemeteries throughout the peninsula; approximately 9,000 of these contain the remains of soldiers who have been identified. Thirteen thousand of the dead are buried in what the CWGC refers to as "concentration" cemeteries, near the locations where their remains were found. The remains of more than 14,000 soldiers were never found; they are commemorated by name on the Helles Memorial (British, Australian, and Indian), the Lone Pine Memorial (Australian and New Zealand), the Twelve Tree Copse, Hill 60, and Chunuk Bair Memorials

(New Zealand).[10] The French troops who lost their lives fighting at Gallipoli also had several burial sites on the peninsula during the war, including a cemetery in the Ottoman fortress at Seddülbahir, but these were concentrated after the war into a larger cemetery at Morto Bay, with burials in both individual graves and ossuaries.

The Ottomans also buried their dead during the war, and they created cemeteries and memorial structures both on and off the peninsula at that time. Some of these were makeshift structures which have been lost, destroyed, or heavily reconstructed. Many of the Ottoman soldiers who perished while fighting in the Çanakkale Wars were buried in mass graves which are still being discovered today with the help of archival military maps and non-invasive archaeological technologies. A more substantial national campaign to commemorate the Ottomans who had died while fighting on the peninsula in the Çanakkale Wars began two decades after the Turkish War of Independence with the submission of plans for the Çanakkale Martyrs Memorial in 1944, an immense commemorative complex referred to today as the Abide (Figure 1.2).

Just before that campaign, in the 1930s, the migration of Turks from the Balkan regions of eastern Europe into western Thrace, which had begun in the late nineteenth century, accelerated as the result of the relative stability in a region

Figure 1.2 The Çanakkale Martyrs Memorial (*Abide*).

Credit: Archive of L. Thys-Şenocak

that had seen so much demographic displacement and upheaval during the Balkan Wars, World War One, and the Turkish War of Independence. Turkish refugees, the majority from Romania and Bulgaria, continued to arrive in the new lands of the Turkish Republic through the 1940s. Many from Romania repopulated the villages on the Gallipoli peninsula, while the majority of the Bulgarian Turks settled in the villages of the opposite Asian shore.[11] Most of these new migrants survived by farming among the makeshift graves of the dead who were buried in their fields, and by selling the salvaged ordnance that remained from the war. Many built their first homes on the peninsula by harvesting the crumbling masonry from nearby Ottoman or World War One military structures, and by melting down ordnance left behind from the war.[12]

In the decades that followed, proposals were made to create a national historical park on the Gallipoli peninsula, but it would not acquire that official status until 2 November 1973. The groundwork had been laid for the formation of national parks in Turkey in the mid–1950s with the establishment of non-governmental organizations such as the Foundation for Turkish Nature Conservation (TTKD) in 1955, and the approval by the Turkish government of the 1951 Agreement on the Establishment of European and Mediterranean Plant Protection Organization. By the 1970s, the environment and its protection had become an official part of Turkey's 1973–78 Five Year Development Plan.[13] The formation of Turkey's first national parks was the result of collaboration in the late 1960s between the United States National Park Service (USNPS) and the State Planning Organization in Turkey, under the direction of the head of the Turkish National Parks Department, M. Zekai Bayer. This project, a USNPS and United States Agency for International Development (USAID) effort, was intended to create plans for twelve national parks in Turkey. Initially Gallipoli was to be among these first parks, along with Troy, but that proposal was rejected by the Turkish park planners, one who proposed that the Gallipoli park be named the National Park of European Aggression.[14] While Gallipoli would become a national park in 1973, Troy and its environs waited until 1996 for that status; the latter site was listed soon thereafter as a World Heritage site in 1998.

From the 1970s through the 1990s a series of architectural competitions was held by different ministries of the Turkish government to create several commemorative projects for the Gallipoli National Historical Park.[15] While the Çanakkale Martyrs Memorial of the 1940s was an immense modernist concrete structure that dominates the southern end of the peninsula, several of the projects generated in the architectural competitions from the 1970s through the 1990s purported to be inspired by a desire to have minimal impact on the landscape, with limited intervention. Architects and jury members participating in these competitions agreed that winning designs should approach Gallipoli as if it were an open-air museum. The results of most of these projects did not have that effect and ultimately were often highly invasive.

An example of how the rhetoric of non-intervention did not match the reality is evident in Ahmet Gülgönen's Conkbayırı Mehmetçik Park epigraphic monuments designed in the 1970s (Figure 1.3). The five large concrete slabs erected at a strategic location on the Gallipoli battlefields were allegedly arranged to resemble a hand reaching from the earth to the heavens in a gesture of prayer. The slabs narrate different stories about the Ottoman soldiers of the 57th regiment and their leader Mustafa Kemal Pasha (Atatürk) who fought for the highly contested ridge of Conkbayırı in this section of the peninsula throughout the spring and summer of 1915.[16] Three of the five texts inscribed on the slabs celebrate an action, order, or story of a fortuitous event that occurred during the battles in which Mustafa Kemal partipated, among these the famous command to his regiment, "I do not order you to attack, I order you to die. In the time which passes until we die, other troops can take our places and other commanders can master the situation." Another describes a miraculous moment when a pocket-watch saved Mustafa Kemal Pasha's life by absorbing the impact of shrapnel or a bullet that was fired at his heart.

The winning project was praised by the jurors for its non-invasive, even subtle approach to the landscape, its evocation of the Orhan inscriptions (the earliest examples of Turkic writing in Central Asia on monoliths from the sixth century), its reasonable cost of construction and upkeep, and for approaching the entire landscape as if it were an open-air museum.[17] While one juror praised the epigraphic monuments for treating the surrounding landscape as sensitively as if

Figure 1.3 One of the epigraphic monuments created by Ahmet Gülgönen at the Conkbayırı Mehmetçik Park.

Credit: Archive of L. Thys-Şenocak

it were a museum, another lauded the monuments for being human in scale. Gülgönen's monuments do, in fact, function like museum labels, but are immense and highly didactic. Their larger-than-life size intimidates the viewer, making any journey of self-discovery or quiet reflection among the slabs rather difficult. Today Gülgönen's commemorative monuments stand as testimony to both our changing ideas about landscapes and the purposes of museums. They also serve as concrete lieux de mémoire of how politics, memory, and architecture have often allied at the expense of landscape.

The destruction of 4,049 hectares of forest in the Arıburnu region during a massive fire on the peninsula on 25 July 1994 was the catalyst for a government-led initiative to create a comprehensive inventory of the peninsula's natural and cultural resources. This project was led in 1997 by Raci Bademli and a team of urban planners from the Middle East Technical University (METU) in Ankara. In addition to conducting research about the Gallipoli peninsula and undertaking a formidable documentation process, Bademli advocated for an end to the wave of development that had occurred in the 1980s and which had resulted in the construction of several symbolic Turkish cemeteries and more commemorative monuments. He proposed that the region become an international Peace Park and that projects developed for the peninsula in the future be less invasive and less destructive of the landscape. The international Gallipoli Peninsula Peace Park Ideas and Design Competition was held the next year; following this a comprehensive inventory and management plan for the existing natural and cultural assets on the peninsula were undertaken by the METU team and published in 2002. The 1998 competition was conceptualized as part of a solution to restructure the national park and its management, and to create non-invasive visitor centers and commemorative routes through the landscape to highlight the importance of the battles fought there; it was also to emphasize the theme of peace, for the benefit of both humans and nature. The competition was won by the Norwegian architects Lasse Brøgger and Anne-Stine Reine, but their project, which was awarded first prize because it called for minimal intervention in the landscape, was never implemented.[18]

Interventions in the Gallipoli landscape in recent decades have not been limited to the Turkish government. In 1999, a year after the Peace Park competition, the Australian and New Zealand governments, citing an increased interest among their citizens in attending the ANZAC Day commemorative ceremony, proposed to the Turkish government the construction of an ANZAC commemorative site. This new site, they hoped, would lessen the impact of the growing crowds attending ceremonial events on the fragile coast where the Arıburnu cemetery is located.[19] The Australian and New Zealand governments pledged 1.2 million dollars to establish a new ceremonial site 300 meters to the north of the Arıburnu cemetery, with the Turkish government committing 100 million dollars to major improvements to the infrastructure (Figure 1.4). An official request from the Australian Minister for Veterans' Affairs, Danna Vale, to the Turkish Minister for Environment and Forests, Osman Pepe, asked for turnaround facilities for larger buses along the coastal road, as well as the stabilization and widening of the road along the coast, the roads from Lone Pine to Chunuk Bair (Conkbayırı), and the Kemalyeri Memorial.[20] What began as a cooperative project among these three countries turned into an international debacle, with all sides accusing the others of negligence and inept planning. The project resulted in major interventions in the landscape and the destruction of coastal areas in the northern Gallipoli region, along with the loss and destruction of cultural heritage, including human remains from World War One. The report from the Australian parliamentary inquiry stated among its findings that it was largely at fault. "The Government, by its failure to safeguard the ANZAC Cove site and its Australian military heritage, defaulted on its responsibility to the Australian people, particularly the expectations of veterans."[21]

With the 2002 election in Turkey of the Justice and Development (*Adalet ve Kalkınma*) party, the dominant narrative about the Gallipoli peninsula, which had developed shortly after World War One and celebrated the role of the nation's founder, Atatürk, in the Ottoman victories on the peninsula, began to change. Veysel Eroğlu, the Minister of Forestry and Water Affairs, stated that a visit by the then Prime Minister (now President) Erdoğan on 18 March 2003 and his orders to "save the martyrs' cemeteries in Gallipoli from the state of neglect" were catalysts

Figure 1.4 The ANZAC commemorative site on the northern shores of the Gallipoli peninsula: preparations for ANZAC Day, 25 April 2014.

Credit: Archive of L. Thys-Şenocak

for major changes to be made on the peninsula. Yet another wave of commemorative building projects was then initiated by the government in 2005, beginning with a visitor's center at Kilye at the entrance to the national park. Based on the maps of the Gallipoli peninsula produced by Şevki Pasha in 1916, several newly discovered Turkish collective cemeteries were marked with symbolic gravestones and planning began to make these sites more accessible to tourists with the creation of more parking lots and wider roads for tourist buses.

As part of the process of reorganizing the peninsula and creating a new narrative for it, the Ministry of Forest and Water Affairs tore down the Kabatepe museum and visitors' center, a structure which itself had won first prize in an earlier 1986 architectural competition.[22] A new architectural competition was organized by the ministry in 2009 for the purpose of creating a simulation and information center and a museum on that same site, in spite of protests from the Turkish Chamber of Architects that the competition was illegal for several reasons—among these, the presence of only one architect on the jury, and the fact that this new project disregarded the results of the earlier 1997–8 international competition which had envisioned the peninsula as a "peace park," with a mandate to intervene as little as possible in the landscape of Gallipoli. With a construction site covering 8,600 square meters, the Çanakkale Destanı Tanıtım Merkezi, or the Legend of

Figure 1.5 The entrance to the Legend of Gallipoli Simulation and Information Center, opened in 2012.

Credit: Archive of L. Thys-Şenocak

Gallipoli Simulation and Information Center, opened in June 2012 (Figure 1.5). At a cost of roughly thirty-three million dollars, the edu-tainment complex also functions as a museum and now redisplays the wartime objects and memorabilia from the earlier Kabatepe museum collection, the private Saim Mutlu museum in Alçıtepe, and objects from other private and state collections.[23]

Activities to commemorate the fallen at Gallipoli through the construction of symbolic memorials and new cemeteries, and to provide tourists with facilities such as the Legend of Gallipoli, continued at a rapid pace under the Turkish Ministry of Forestry and Water Affairs until the peninsula was removed from that ministry's jurisdiction in 2014.[24] The results of a decade (2005–14) of development of the landscape of Gallipoli, and the emergence of a robust faith tourism, have now fallen to the Turkish Ministry of Culture and Tourism's Çanakkale Wars and Gallipoli Historic Region Directorate (ÇATAB) to manage. How it will navigate the interests that private individuals and corporations have—such as OPET, a petroleum company—in contributing to corporate branding through social responsibility projects, and the "business" of commemoration on the peninsula, is yet another challenge.[25]

In 2014, along with the change in administrative control of the Gallipoli peninsula from the Ministry of Forestry and Water Affairs to ÇATAB and the Ministry of Culture and Tourism, the Gallipoli peninsula lost its status as a national park. In addition to the restructuring of the peninsula under the direction of ÇATAB and the initiation of a long-term development plan targeted for completion in 2023 (the centenary of the founding of the Turkish Republic), Turkey's 2014 submission of the Gallipoli peninsula to UNESCO for consideration as a World Heritage site was among the most significant developments to have occurred on the peninsula in the past decade. This nomination reflects Turkey's growing interest in promoting Gallipoli as a shared heritage place of universal significance based on the events that occurred there in World War One.[26] The strategies the new ÇATAB directorate is now using to manage the landscape of Gallipoli and its many stakeholders, and its plans for the next decade, are examined in more detail in the final chapter of this book, but it is useful to introduce here some information about current activities and their impact on the shifting identities of the peninsula.

Changing narratives, challenging identities

For three of the countries that participated in the Gallipoli campaign—Australia, New Zealand, and Turkey—the peninsula and the events that happened there between 1914 and 1916 are regarded as formative in their founding narratives. For Australians and New Zealanders, the ill-fated landing at dawn on 25 April 1915 on the beaches of Arıburnu and the ensuing bravery of the ANZAC troops in the face of daunting topographic and military obstacles are a particularly integral part of the present-day narratives these two countries have about early statehood formation.[27] For Turkey, some scholars working on the early years of the Republic have noted, "Gallipoli became so much identified with the national foundation myth that people often mistake it as a part of the War of Independence."[28]

For many Turks the landscape of the Gallipoli peninsula has been associated with the heroism and tenacity demonstrated by the nation's founder, Mustafa Kemal Pasha, or Atatürk, in the face of a formidable adversary: Allied Europe. Since the signing of the Treaty of Lausanne in 1923, in which the European powers returned the lands on the Gallipoli peninsula to the Turks with a proviso to keep limited jurisdiction over their World War One cemeteries, these battlegrounds have also been linked to other aspects of Atatürk's later achievements: the struggles to create a secularist nation, to preserve the territorial integrity of the land that comprises modern-day Turkey, and to promote Kemalism, the political "ism" that encapsulates the vision that Atatürk had for the future of Turkey.[29] Confident of the enduring power of the Turkish Republic's founding narrative and the role played by Mustafa Kemal Pasha in the Çanakkale Wars, Vedica Kant has argued that "despite greater acknowledgment of ordinary soldiers, and a more substantive challenge to the Kemalist narrative by the Islamic Justice and Development Party (AKP), that myth has proved deeply resilient, limiting the terms in which the significance of Çanakkale can be contested and reoriented in Turkey."[30]

However, since 2005, the significance of the Gallipoli peninsula and the role of Atatürk in these narratives has changed. With a few exceptions, what was a largely secularist and nationalist-oriented narrative and tourist itinerary dedicated to celebrating many of the military triumphs of Atatürk during the war has been gradually recast to reflect a more faith-based narrative with an itinerary in which religious rituals are integrated into the nationalist pilgrimage.[31] When politically advantageous, this faith-based narrative embraces the peoples and religions of all who fought with the Ottomans, including Armenians, Alevis, Arabs, Kurds, and Jews. But the narrative can quickly change to exclude minority groups. The performative aspects of this narrative have generated pilgrimage routes and spaces of prayer which highlight the sacrifices made collectively by Muslim martyrs to protect their country. Superstitious beliefs, tales of miracles, and stories of the collective sacrifices of devout Muslim martyrs have been integrated more frequently and enthusiastically into the explanations given by some tour guides about Ottoman military successes, occasionally replacing the stories about the heroics, resourcefulness, and leadership of Mustafa Kemal Pasha that had been taught to previous generations. Today, a visit to the Gallipoli peninsula for Turkish schoolchildren, or for newly conscripted soldiers, could follow a somewhat different path and tell a different story from the one familiar to their parents and grandparents. Some have even remarked that the identity of the peninsula and a visit to it is essential for Turks as it constitutes a form of the Haj.[32]

The Gallipoli landscape as androcentric: the role of gender and conflict in archaeology

Along with these shifting identities and narratives about the peninsula that the tourist or pilgrim may discover, there has been a growing awareness, especially among archaeologists working there, that there is relatively little known about this landscape, particularly considering the long history of archaeological excavations on the opposite shores of the Dardanelles at sites such as Troy. Because of the centrality of the Çanakkale Wars in the historical narratives about Gallipoli, the discourse about its landscape has also been largely androcentric; but new approaches to archaeological work, which are non-invasive, or which bring different perspectives such as gender or the analysis of conflict to both the practice of excavation and the interpretation of finds, have the potential to bring fresh insights to the archaeology conducted on the Gallipoli peninsula.[33]

There has been a very long history and deep stratigraphy in which women were present on the peninsula—in its mythology, temples, homes, villages, farms, and hospitals—yet the primary identification of Gallipoli forefronts the time of the Çanakkale Wars and the men who fought, died, and were buried in the earth of the peninsula during and after those wars. While "adding women and stirring" is now considered by many archaeologists to be an outdated approach to rectifying a gender-biased discourse about landscape, the absence of women in most narratives about the geographies of Gallipoli is striking.[34] There has been a substantial amount of research conducted in the past decade on the diverse roles played by both Allied and Ottoman women in the Gallipoli battles, on and behind the front lines, yet the issue of gender in the Çanakkale Wars has also been of little concern in the museum exhibitions or tourist narratives about the peninsula. Women make occasional appearances as mothers of soldiers, mourners of the dead, and less frequently as nurses. Considering that Helle was the eponymous female founder of the Hellespont which formed the European border of the peninsula; that major fortifications on the peninsula were constructed and repaired by the mother of an Ottoman sultan, Ottoman women snipers fought on the front line during the Gallipoli battles, and female nurses tended to the wounded as their consoeurs supplied the front through work in the fields and the factories of the empire, the absence of women and the larger issue of gender and how it intersects with Gallipoli's landscape should be addressed. Archaeology may be one way to cut through the layers of Gallipoli's past which have largely excluded women, challenge the assumptions that have been made, and pose new questions about the gendering of the Gallipoli landscape through time.[35]

There are in fact many stories embedded in the lands of the Gallipoli peninsula and the waters that surround it; stories that occurred prior to, during, and after the two years when this long spit of land was catapulted to fame during World War One. Each layer is an important component of the heritage of this place. While the story of Gallipoli during the Çanakkale Wars is the most dominant narrative associated with this landscape, it is essential to recognize the heritage values of the other layers if they are to be preserved. Who will shape the narratives about Gallipoli's past, and how the many heritage(s) of Gallipoli have been, and will be, interpreted in the future is the central concern of this book.

The Gallipoli peninsula as heritage

While acknowledging the immense importance of the battles fought, and the many lives lost on the Gallipoli peninsula during World War One, *Divided Spaces, Contested Pasts: The Heritage of the Gallipoli Peninsula* is not a military history of the Gallipoli campaign. There are many detailed accounts of the campaign which do focus on the strategies, alliances, and various aspects of the battles fought on the peninsula, from both the Ottoman and European perspectives. Particularly since the 2015 centenary, an abundance of histories and detailed analyses has been written about the combined naval and ground battles fought on the Gallipoli peninsula and in the waters of the Dardanelles between 3 November 1914 and 9 January 1916. The reader interested in those aspects of Gallipoli's military history will find no shortage of good primary and secondary sources.[36] This book, however, focuses on the heritage of the peninsula after World War One, and more specifically on how and why this iconic cultural landscape has become a place of contested heritage, both in Turkey and internationally.

Looking at the most recent century of the peninsula's past through the lens of heritage, *Divided Spaces, Contested Pasts* investigates how different pasts and meanings have been constructed for and about the peninsula in the past century to accommodate a range of commemorative needs and economic and political agendas. By approaching the landscape of the post-war Gallipoli peninsula from a heritage perspective, a major concern of this book is how interpretations of this landscape's past are shaping future plans for this region, how it will be managed in the coming decades, and who will have physical and intellectual access to it. As Peter Howard has pointed out, what differentiates the older discipline of history from its postmodern and more practically oriented offspring is that while "history is interested in the past, heritage is interested in how the past might be conserved and interpreted for the benefit of the present and the future."[37]

Approaching a site like the Gallipoli peninsula from the perpective of heritage recognizes that whatever meanings are attached to that landscape are part of a changing and dynamic process in which a variety of peoples and communities (commonly referred to as "stakeholders") are involved in the construction of socially-produced space and its heritage(s).[38] The methodologies used in history and critical heritage studies allow for the existence of different, and frequently contradictory narratives and even encourage different interpretations to exist and to be contested.[39] As the title of this book suggests, I believe that several different narratives and interpretations about the pasts of the Gallipoli peninsula have developed in the past century, and that the landscape of the peninsula is divided, both conceptually and physically, into spaces with multiple meanings and identities. If the past century gives us any indication of what we can expect in the future, it is likely that the Gallipoli peninsula will continue to serve new and competing agendas in the coming century. It is also likely that each administrative body that has jurisdiction over the peninsula will want to leave some tangible legacy of its presence on this landscape, as this is what has happened in the past, particularly through the construction of commemorative monuments. Recognizing the ways in which the different heritage(s) of the peninsula have been neglected, manipulated, constructed, memorialized, and forgotten since World War One by a wide range of local, national and international stakeholders is crucial at this time because of the continual, tangible, and frequently destructive interventions in the cultural landscape of Gallipoli fueled by changing political and economic agendas.

The Gallipoli peninsula as World Heritage

Three recent developments have made it particularly important to examine the peninsula as heritage. The first is the aforementioned proposal, made on 15 April 2014 by Turkey's UNESCO office, to nominate the "Çanakkale (Dardanelles) and Gelibolu (Gallipoli) Battles Zones in the First World War" to the UNESCO tentative list of World Heritage sites.[40] The second development, announced in the official gazetteer of the Turkish Parliament on 19 June 2014, is that the "Gallipoli Peninsula battlefield places would begin a new era" without the national park status gained in 1973. Instead, the Gallipoli peninsula (while still required to follow national legislation related to the protection of the natural and cultural assets of Turkey) is now administered by ÇATAB, the aforementioned Çanakkale Wars and Gallipoli Historic Region Directorate.[41] Third, with the passing of the centenary of World War One, much of the material remains from that war, including the maritime heritage in the waters surrounding the peninsula, could come under more robust national and international heritage legislation, such as UNESCO's Convention on the Protection of the Underwater Cultural Heritage.[42]

Nomination of a heritage site to the tentative list of UNESCO World Heritage sites is the first step required for permanent inscription to the list of World Heritage Organization sites. The selection and nomination of a site to the tentative list must be suggested to UNESCO and approved by the state party submitting the nomination, in this case Turkey. Therefore the proposal that the Gallipoli peninsula should be added to the World Heritage tentative list is an official and public acknowledgment by the government of Turkey that it conceptualizes the landscape of the Gallipoli peninsula to be a place of outstanding universal value that deserves recognition for its significance as world heritage. To be considered by UNESCO for full inscription to the World Heritage List, at least one of six criteria defining outstanding and universal heritage values of a site must be met.[43] The Gallipoli peninsula was nominated under "criterion 6" in its inscription dossier, as it was said to be "directly or tangibly associated with events or living traditions, with ideas, or with beliefs, with artistic and literary works of outstanding universal significance."[44] The nomination file created for Gallipoli describes the role that the peninsula has played both as a cultural bridge and as a barrier since Neolithic times, but stresses that ultimately this landscape is important for more nations than just Turkey because the "Çanakkale and Gallipoli battles constitute a landmark in the world military and political history."[45]

The nomination of the Gallipoli peninsula to the tentative list of World Heritage sites in 2014 by the Turkish government was not the first time that the peninsula had been proposed and promoted as heritage. Australian Prime Minister John

Howard proclaimed in 2003 that the ANZAC Cove of Gallipoli should represent the first nomination for inclusion in the National Heritage List of Australia.[46] Records of the proceedings of the Legislative Assembly of New South Wales from July 2003 reveal that this nomination was a popular idea, at least among some government officials in Australia.[47] However, most professionals and academics in the heritage field spoke out against the proposal and were highly critical of Australia laying claim to a national heritage site in a foreign country. Lynn Meskell has described the suggestion by Howard and his supporters as a manifestation of the weakness of the Australian government to deal with the internal politics of the country and remarked that, "Claiming heritage in a foreign conflict zone has clearly proven more palatable to the Australian government than addressing its own internal repressions and seeking equitable restitution for segments of its citizenry."[48]

However, the efforts of the Australian prime minister to assert a more tangible claim to sections of the Gallipoli peninsula did resonate to a certain degree among the general public, as noted by war historian John McQuilton who writes, "there is little doubt that most Australians and New Zealanders see Anzac Cove on the peninsula as 'theirs' . . . the peninsula may well be in a foreign country but 'ownership' was somehow conferred by the loss of Australian and New Zealanders' lives."[49] Some government representatives, such as former Australian Veterans' Affairs Minister Danna Vale (who was involved in the campaign discussed earlier to widen the roads at Arıburnu), even proposed solving the problem of ownership of the heritage of the Gallipoli peninsula by creating a new ANZAC memorial site in Australia. Vale claimed that the resemblance of the Mornington peninsula in Victoria was close enough to the Gallipoli landscape to serve as a substitute, and would save many of her compatriots from making the long and costly journey to Turkey.[50]

Uneasiness between Turkey and Australia over the legal ownership of the actual land upon which the ANZAC graves are located has existed since the signing of the Treaty of Lausanne in 1923, in which Articles 124–136 describe the rights and restrictions that different signatories to this treaty have regarding their gravesites on the peninsula.[51] Concerns over any changes made (particularly to the northern section of the peninsula where the majority of the ANZAC graves are located), such as the widening of roads or construction of Turkish memorials, are quickly reported in the Australian press, a reflection that the Australian conceptualization of the peninsula, along with the routes that its citizens follow when visiting Gallipoli and commemorating their dead, prioritizes the ANZAC region of the peninsula over all others, as this is where the majority of its troops fought and died. Maps of the Gallipoli peninsula that appear in some Australian guidebooks and even parliamentary reports frequently show only the limited ANZAC region of the peninsula. Some Australians and New Zealanders when visiting the peninsula never venture beyond the ANZAC area or make the twelve-kilometer journey to the Cape Helles region where there was also ANZAC participation and loss of life, but far less than in the northern section of the peninsula.[52]

Today, within the ANZAC sector itself, the Turkish and Commonwealth graves are clearly defined by different memorials and events, and the narratives which have been created for these have the strongest voices on the peninsula; many are adjacent to each other, as were their trenches during the war. One popular guidebook has noted that the statues of Turkish war heroes and the Commonwealth obelisks occasionally appear to be "locked in parallel monologues."[53] While these two monologues have been the loudest on the peninsula, in the past decade there has been a growing interest in remembering other voices, communities, and nations which were part of the World War One past on the Gallipoli peninsula.[54] Previously "forgotten peoples," such as the indigenous tribes of New Zealand who were part of the ANZAC troops, are now laying claim to commemorative space and time in the annual 25 April ANZAC ceremonies. In the past decade the French, British, Irish, and Russians have also shown more interest in remembering their losses on the Gallipoli peninsula during the war, and in marking those with commemorative acts and architecture.[55]

The Gallipoli peninsula and material culture: human remains, ordnance, and World War One artifacts

Since the centenary, material culture on the peninsula from World War One is also taking on a new meaning and legal status. The removal of both archaeological objects and other types of material culture, including World War One ordnance and memorabilia, from the peninsula is not a new phenomenon. Many of the archaeological finds from Elaious and Karaağaçtepe, the so-called Protosilaus mound, which were excavated by the French during and shortly after the war, are now in the Louvre. Shiploads of World War One ordnance (described as "trophies") were salvaged from the Gallipoli peninsula in 1919 by the Australian historian C.E. Bean and his team to establish the national collections in Australian museums.

A new awareness of the value and significance of artifacts and finds from World War One, including human remains and ordnance, has emerged only recently in Turkey. The need for official procedures and protocols concerning the excavation, conservation, management, and display of these types of finds has only just been addressed in 2017 in the ÇATAB excavation at the Seddülbahir fortress, where archaeological human remains of Ottoman and Allied soldiers were discovered along with various types of potentially explosive ordnance. In several of the museums and private war galleries on the peninsula, both types of finds can still be seen in display cabinets. Equally as important as issues of design, exhibition content, and interpretive strategies for the World War One collections in the museums on the peninsula is the fact that many of the objects in these museums and private collections still need to be evaluated both for safety issues (in the case of the display of wartime ordnance) and for the ethical implications of displaying archaeological human remains, particularly those from World War One. The need for official policies, standards, and guidelines for the excavation and management of archaeological human remains from all eras is only now being debated in Turkey, and the results of these debates will have significant implications for archaeologists, museum curators, and anyone who lives and works with the material culture that is discovered in the land and seascapes of Gallipoli, from farmers and fishermen to contractors and tour operators.[56]

Increased state protection of World War One material culture, and legislation regulating its ownership, are also impacting the Gallipoli peninsula. Personal collections of these types of objects owned by residents on the peninsula are now subject to confiscation by the state if an official collector's license is not obtained; if found with recently designated historic objects, the unlicensed collector can be fined, imprisoned, or both.[57] Unfortunately, as the historic and cultural value of World War One objects has increased, so has the theft of material culture of this era from the peninsula's battlefields, museums, and underwater heritage sites, such as the wrecks of warships, barges and submarines off its shores. Treasure hunting in general has accelerated in the past decade as the market value and demand for these types of items have increased, both in Turkey and abroad. Promoting museums and centers for the study and scientific practice of World War One archaeology (as has been done in Europe by scholars such as Nicholas Saunders), and adhering to international policies and practices concerning the management of historic battlefields of World War One and underwater archaeological heritage (such as the Vimy Declaration for the Conservation of Battlefield Terrain, and the 2001 UNESCO Convention on the Protection of the Underwater Cultural Heritage) are just a few of the steps which could be helpful in managing the very diverse archaeological heritage of the Gallipoli peninsula.

The Gallipoli peninsula as a cultural landscape

Within the larger category of heritage, the physical landscape of the Gallipoli peninsula is considered and analyzed here using a heritage instrument known as a "cultural landscape."[58] While this was not the heritage instrument that the Turkish Ministry of Culture and Tourism utilized in 2014 when proposing the peninsula for the World Heritage tentative list, the term was referred to in the planning workshop organized by ÇATAB in February 2017 to describe the landscape of Gallipoli. The 2016–2020 strategic plan for the Gallipoli peninsula produced by ÇATAB also stresses the importance of developing the region as an open-air museum with ecologically sustainable practices that will balance the needs of residents, visitors, and the landscape.[59]

Both the terms "cultural" and "landscape" are rich in definitions, many of which have changed significantly in the past three decades. Carl Sauer (1889–1975), who is often considered to be the first cultural geographer, is frequently credited with first putting together the words "cultural" and "landscape." Landscape, he claimed, is "a way of seeing that has its own history . . . that has its own assumptions and consequences . . . whose origins and implications extend well beyond the use and perception of lands . . ."[60] Interpreted in this way, landscape is far more than just a place to look at, as it has the potential to carry diverse political and social agendas. More than a "transparent window through which reality may be viewed unproblematically, the cultural landscape is now conceived of as an 'emblematic site of representation, a locus of both power and resistance.' "[61]

The concepts embedded in the category of cultural landscape owe much of their theoretical underpinnings to intellectual developments after the 1980s, and the impact of the "cultural turn" in the fields of landscape studies, park and environmental history, cultural studies, and art history. Particularly after 1980, new approaches to the symbolics of landscapes and the encouragement of multivalent interpretations of heritage sites demonstrated that cultural landscapes constituted spaces which were, as Henri Lefebvre has written, "permeated with social relations . . . but also producing and produced by social relations."[62] In landscape studies, authors such as Denis Cosgrove and Stephen Daniels have pointed out that landscapes need to be deconstructed and re-read in much the same way that "new historians" reassess their documents, mining them now not only for the alleged facts but also to determine the mindsets and biases of the individual and collective memories that are created and reflected in the texts themselves.[63]

Cultural landscapes, particularly those associated with World War One and the memories that have been inscribed on landscapes associated with war, have also been among the subjects of Pierre Nora's extensive work on the collective memories of France, and his envisioning of these places as lieux de mémoire.[64] As so much of the land on the Gallipoli peninsula is associated with the collective memories of the dead from so many different countries, Nora's work is helpful when understanding Gallipoli's ability to shape the imagination and memories of many who have never set foot on the peninsula. What Jay Winter claimed for World War One battlefields of the Western Front in Europe rings equally true for Gallipoli: "Remembrance is part of the landscape."[65]

Employing new frameworks to analyze and categorize heritage sites brings opportunities but also difficulties. One of the challenges of using cultural landscapes as a way to understand the heritage of Gallipoli is the risk of this category being too large, cumbersome, and inclusive to be useful. Several scholars have weighed in on this challenge. At one extreme, environmental groups report that virtually every place on earth has been altered by the human presence and there is nothing left which can be considered "pristine"; from this assumption we could deduce that virtually every space can be defined as a cultural landscape.[66] In *Landscape and Memory*, Simon Schama argues that "landscapes are culture before they are nature; constructs of the imagination projected onto wood and water and rock."[67] Peter Howard has also remarked on the large number of academic disciplines now using the concept of the cultural landscape, and finds this attraction to the category by so many different users as one of its strengths rather than a weakness: "if it had become the intellectual property of a single discipline it would by now have been defined so carefully as to miss all the richness of the present confusion. . . ."[68] Acknowledging the hazards of using this category, and its potential to encompass too much, it is precisely because of its flexible and multidisciplinary appeal that it presents an appropriate framework in which to situate and analyze the heritage of a place as complex as Gallipoli. [69]

"Cultural landscape" became an official category to designate heritage sites in 1992 at the World Heritage Convention; additional definitions of it were later articulated in the European Landscape Convention, which was adopted in 2000 and activated in 2004. In these contexts a "cultural landscape" was conceptualized as an instrument that could define particular types of heritage places where a separation of natural and cultural assets did not make sense. The World Heritage Committee decreed that cultural landscapes are, in fact, the "combined works of nature and man" and are "illustrative of the evolution of human society and settlement over time, under the influence of the physical constraints and/or opportunities presented by their natural environment and of successive social, economic and cultural forces, both external and internal."[70] The underlying assumption made by UNESCO and others working with the concept of the cultural landscape in the heritage sector is that "nature" and "culture" should not be conceptualized or preserved as separate entities. Rather, nature and culture are seen as inextricably intertwined and it is the interaction between the two that is important to recognize and value. In other words, landscapes and the different meanings inscribed onto them do not just exist in a pristine vacuum; they are continuously changed, shaped, and reshaped by humans in response to natural as well as political and economic forces.

However, recognizing and categorizing this interaction and dependency between the landscape and those who live within it should not result in uncontrolled human intervention at the expense of that landscape. Ultimately cultural landscapes are spaces where there is a complex web of relationships between humans and the land; humans don't just impact, but also interact with that landscape in the production of both culture and heritage. They are, as Lowenthal has remarked, "Creatures of historical processes beyond our control, we shape landscapes and artifacts to conform with illusory histories, public and private, that gratify our tastes. . . ."[71] If this is the case we must be cautious not to reshape these beyond recognition and cause irreparable loss of all that allowed us to define and value that landscape as heritage.

The Gallipoli peninsula: divided and contested narratives and spaces

This book makes the claim that the Gallipoli peninsula is a cultural landscape that is both divided and contested in terms of the discourses and narratives that have been, and continue to be used by different stakeholders about the peninsula. This may be surprising considering the official rhetoric and ceremonies organized by the governments of Australia and Turkey that highlight the battlefield camaraderie that evolved among former foes. The Turkish government has played a major part in the production of the story about the unique relationship between the Turks and Australians as brothers-in-arms during the actual battles, and has given the Gallipoli campaign/Çanakkale Wars the distinctive title of being the "last gentleman's war," a phrase used in the official 2014 Turkish application to nominate the Gallipoli peninsula to the World Heritage Tentative List.[72] Along the major tourist itineraries of the peninsula, iconic sculptures such as the larger-than-life bronze casting of an Ottoman soldier carrying a wounded Allied soldier reinforce this theme of the "gentleman's war"[73] (Figure 1.6). One of the most frequently quoted passages which appears in the literature about Gallipoli, from textbooks to tourist guides, is the message Atatürk sent in 1934 to an Australian delegation visiting the battlefields. On that occasion he assured those who had come to mourn their loved ones that their "sons are now lying in our bosom and are in peace . . . they have become our sons as well."[74] These heartfelt words are inscribed on memorials positioned throughout the Gallipoli peninsula and resonate on the opposite side of the world on commemorative placards and monuments erected by the Australians, such as the Atatürk Memorial in Canberra, Australia.[75]

The film industry has also played a key role in keeping this poignant narrative of camaraderie-in-conflict alive. Filmed in other parts of Turkey such as Cappadocia, where the barren treeless landscape more closely resembles that of the battle-scarred Gallipoli of World War One, these movies have generated

Figure 1.6 "The Deep Respect for Mehmetçik (Turkish Soldier's) Memorial" bronze sculpture shows an Ottoman soldier carrying a wounded Allied soldier back to the safety of his trenches.

Credit: Archive of L. Thys-Şenocak

fame for this now heavily reforested corner of Turkey. Peter Weir's production of "Gallipoli" in 1981 starring a young Mel Gibson told us about "a place we've never heard of and a story we'll never forget." Russell Crowe's film *The Water Diviner* (in Turkish "Son Umut," The Last Hope) was released on 26 December 2014. Here the main character, Australian farmer Joshua Connor (played by Crowe), leaves the outback of Australia to search for the remains of his three sons whom he believes are buried on the beaches of Gallipoli. During his quest, and as testimony of deep friendship and respect between former foes, Connor ends up saving the life of his Turkish co-star, who is fighting against the Greeks in the Turkish War of Independence. He then falls in love with a Turkish war widow, and they live happily ever after in Istanbul.[76]

The mythology of the "gentleman's war" has provided much grist for the mill for Turkish and other novelists writing about Gallipoli during and after the war. Azer Kemaloğlu has described the way the history of the conflicts on the peninsula has served as a "dialogic medium to activate fictional representations"; and this is evident in several Turkish novels, among these Buket Uzuner's *The Long White Cloud: Gallipoli* (2001) and Eyüphan Erkul's *Savaş Ana ("Mother War")* (2014) in which the authors demonized war itself but idealize the intimate relationships that form between those who are supposed to be enemies.[77]

The reality of the relationships that have developed over the past century between members from the former Allied countries and the Turks who inherited the peninsula from their Ottoman ancestors is more complicated, both in terms of how the diverse narratives promoted about the peninsula have evolved, and the ways in which the actual space of the peninsula has been defined by commemorative building projects or delineated for different commemorative acts, pilgrimage and tourist routes, political and cultural events. In some cases the divisions, access to, and ownership of specific geographies of the peninsula have been created by official decree, as with the aforementioned Treaty of Lausanne.[78] Particularly in Turkey and Greece, any questioning of this international agreement, now close to a century old, still sends political shockwaves through the Aegean and beyond. But in most cases the divisions of the peninsula have emerged over time and through a complex web of ritual, memory, conflict and compromise.

The division of the peninsula into separate and contested geographies has been a constant phenomenon which continued long after World War One ended; the contested narratives and claims to heritage go beyond territorial debates between Turks and Greeks or Turks and Australians over the interpretations of a treaty signed in 1923. How to commemorate the heritage of Gallipoli has been, on occasion, a source of division between the British and Irish, British and French, Australians and New Zealanders, Aboriginal Peoples and Australians, secular and more religious Turks, residents of the peninsula and bureaucrats in Ankara, archeaologists and developers, and peoples of different genders.

Today on the peninsula many residents live with the fear that their homes and fields will be expropriated for projects which have been generated in government offices in Ankara, and about which they were never consulted and have had little say. The 2016–2020 strategic plan generated by ÇATAB stresses the importance of consulting local stakeholders, but there is a long legacy of top-down decision making about the Gallipoli region which has been done by both national and international stakeholders. Consultation with local communities on the peninsula, a cornerstone of contemporary heritage practices, was not consistently practiced by the Turkish Ministry of Forestry and Water Affairs or the sponsors of some of its projects during its administrative tenure on the peninsula, even though a Historical National Park Management Local Consultancy Committee was established to "provide a participant and transparent area management" involving the peninsula's mayors and local village heads (*muhtars*).[79]

Conflicting views about who has a legitimate claim to the heritage on the peninsula, and what identities it should be assigned have appeared in many different forms, from films to "peace parks" projects, to the construction of major architectural monuments. As the commemorative architecture on the peninsula is the most tangible evidence of intervention in the post WWI cultural landscape, and because I came into the heritage field as an architectural historian, I have devoted a section of the last three of the chapters in this book to exploring the planning rationale, iconography, and heritage of a select group of commemorative structures: the Helles Memorial, built in 1921 by Sir John Burnett; the Çanakkale Martyrs Memorial, the construction of which began in 1944 and was designed by Feridun Kip, İsmail Utkular, and Doğan Erginbaş; and the museum, simulation and information center referred to officially as the "Legend of Gallipoli" Simulation and Information Center, an immense edu-tainment complex built in 2012 by the Turkish Ministry of Forest and Water Affairs at the site of Kabatepe, four kilometers south of the Arıburnu commemorative site.[80] The two chapters which immediately follow this introduction explore Gallipoli first as a cultural and then as an archaeological landscape. Chapters 4 and 5 examine the commemorative landscapes of the Allied forces and the Ottomans respectively. My conclusion evaluates the peninsula as a political landscape as Turkey and Gallipoli move towards the 2023 centenary.

Conclusion

Since the resurgence of terrorism in Turkey in 2014, an attempted coup in 2016, and an increasing number of civilian, and military fatalities as a result of internal and external violence, Gallipoli's identity as a landscape dedicated to Muslim martyrs has been strengthened by a state-generated narrative which ties the ambitions of those Ottoman martyrs who gave their lives to protect Turkey a century ago to the goals of those who have died while protecting the present-day Turkish state and preserving its integrity.[81] The tone and intensity of that rhetoric have increased since the attempted coup of 15 July 2016. Bumper stickers on minibuses and fans at football matches now proclaim that "martyrs never die and the homeland will never be divided" as Ankara pulls on the heartstrings of Turkey's citizens and ties them to their collective memories of a quintessential time and place—the Çanakkale Wars and the Gallipoli peninsula—where so many gave their lives to save what would become a cornerstone of the Turkish Republic.[82]

Recently constructed museums and visitor centers, such as the Legend of Gallipoli Simulation and Information Center, the Akbaş Memorial and Monument, Ağadere Hospital and Martyrs complex, (completed in 2015 and never opened), and the 2018 Alçıtepe mosque complex, reflect the present Turkish government's changing agenda for the Gallipoli peninsula. Part of this agenda includes appropriating, and conflating, rather than obliterating the secularist narrative of earlier decades in which Atatürk played a larger role in the identity of the landscape, monuments, and rituals of Gallipoli. ÇATAB's 2016–2020 strategic planning document professes the importance of preserving the peninsula as an open-air museum and establishing an integrated approach to the management of the peninsula which will empower local communities and will protect and conserve historical and archaeological sites, including the underwater archaeological heritage of World War One that lies close to the shores of the peninsula. There is also the promise of conceptualizing and managing the history, archaeology, and heritage of the peninsula within a larger framework that includes the Dardanelles, the islands off the coast of the peninsula, and sites such as Troy, on the opposite shore of the straits. These are ambitious but achievable goals as Turkey and Gallipoli leave behind the centenary of World War One and move towards the 100th anniversary of the Turkish Republic in 2023.

Ultimately the Gallipoli peninsula is a landscape upon which many individual and collective narratives have been inscribed. These different narratives serve to establish, articulate, and explore the memories and identities connected to this place. Many of the interpretations of Gallipoli's past and visions of its future have been and are currently contested, and I believe that there is—and there must be—space for contestation. Some would argue that "difficult heritage" makes for a richer experience of the landscape.[83] I agree, but within that space of contestation the landscape itself must also be able to survive and not be compromised repeatedly for various political and economic agendas. Looking at the Gallipoli peninsula as heritage encourages us to think about how this geography has been reshaped in the past century for so many different purposes; it should also enable us to find effective strategies for preservation and commemoration which are less invasive, which do not destroy irreplaceable natural and archaeological heritage, and which work towards mitigating the polarized narratives which currently divide the spaces, rituals, stakeholders, and agendas on the peninsula into private or state, the religious or secularist, national or international, male or female.

Divided Spaces, Contested Pasts: The Heritage of the Gallipoli Peninsula asks "How much more intervention can the landscape of Gallipoli sustain?" This question is crucial to pose now because the landscape here, while known internationally as a historic battlefield of World War One, is closely associated in Turkey with the founding of the new nation of Turkey. While 2018 marks the conclusion of four years of centenary events commemorating World War One throughout Europe, Turkey is just beginning preparations to celebrate the centenary of the founding of the republic in 1923. Gallipoli will certainly be positioned as a stage for these celebrations in 2023 as it is such an essential geography for the founding myths of this country. By assessing what has happened to the Gallipoli peninsula in the past century, I hope that the complex layers of this fascinating landscape, its heritage(s), and its past(s) can be better understood, and in turn, stewarded carefully into the future. At each turn in my own journey through Gallipoli, David Lowenthal's wise words make more sense to me: "We need the past, in any case, to cope with present landscapes."[84]

Notes

1 C.J. Mackie et al., "Boundary and Divide: the Antiquity of the Dardanelles," in *Anzac Battlefield: A Gallipoli Landscape of War and Memory*, ed. Antonio Sagona et al., (Cambridge: Cambridge University Press, 2016), 5.

2 Lucienne Thys-Şenocak, "Divided Spaces, Contested Pasts: The Gallipoli Peninsula Historical National Park," in *Public Nature: Scenery, History, and Park Design*, ed. Ethan Carr, Shaun Eyring, and Richard Guy Wilson (Charlottesville: University of Virginia Press, 2013), 127.

3 For research on the archaeology of the Gallipoli peninsula and the surveys conducted there since the 1960s see Chapter 3.

4 Mackie et al., 16.

5 Ulrich Wilcken, *Alexander the Great* (New York, W.W. Norton, 1997).

6 Geoffrey Greatrex, "Procopius and Agathias on the Defenses of the Thracian Chersonese," in *Constantinople and Its Hinterland: Papers from the Twenty-Seventh Spring Symposium of Byzantine Studies, Oxford April 1993*, ed. Gilbert Dagron and Cyril Mango (Aldershot: Variorum, 1995), 109–24; Lucienne Thys-Şenocak, *Ottoman Women Builders: The Architectural Patronage of Hadice Turhan Sultan* (Aldershot: Ashgate, 2006), 112.

7 Sarah Midford, *Gallipoli, Anzacs and the Great War* (Melbourne: La Trobe University Press, 2017), 110. https://library.latrobe.edu.au/ebureau/pdf/Gallipoli.pdf.

8 https://nzhistory.govt.nz/media/interactive/gallipoli-casualties-country; the estimates for the number of Ottoman dead vary significantly; this figure is according to the Turkish Ministry of Defense. *Şehitlerimiz* (Our Martyrs); Ankara: 1998).

9 Justin McCarthy, *Muslims and Minorities: The Population of Ottoman Anatolia at the End of the Empire* (New York: New York University Press, 1983), 130–1.

10 These are the statistics provided by the CWGC in the brochure distributed in 2012 at their archives in Maidenhead, England entitled, "The Gallipoli Campaign." n.d.

11 Justin McCarthy, *Population History of the Middle East and the Balkans* (İstanbul İsis Press, 2002), 61.

12 L. Thys-Şenocak and İ. Ceren Cenker, "Local and National Memories of Turkish Fortresses," in *Oral History and Public Memories*, ed. Paula Hamilton and Linda Shopes (Philadelphia: Temple University Press, 2008), 65–86.

13 K. Güçlü and F. Karahan, "A Review: The History of Conservation Programs and Development of the National Parks Concept in Turkey," *Biodiversity and Conservation* 13 (2004): 1373–90.

14 Email correspondence with Hugh Miller, one of the US National Park Service planners who worked on this project (5 March 2009); see also, M. Z. Bayer, *Amerika Birleşik Devletleri'nde Tabiat Kaynakları ve Organizasyonu* (Ankara: Tarım Bakanlığı Orman Genel Müdürlüğü: 1964).

15 For an inventory of the Turkish cemeteries and memorials that were built until 2002 see Raci Bademli, K. Burak Sarı et al. *Gelibolu Yarımadası Tarihi Parkı (Barış Parkı) Uzun Devreli Gelişme Planı Çalışmaları* (Ankara: Middle East Technical University Press, 2002), 75–6.

16 "Conkbayırı Mehmetçik Park Anıtı Proje Yarışması," *Mimarlık* 11 (85) (1970) 34–44.

17 Ibid., 36.

18 R. Bademli, "Gelibolu Yarımadası Barış Parkı Projesi Uluslararası Fikir ve Tasarım Yarışması ve Sonrası," *Mimarlık* 36 (283) (1998), 25–6.

19 www.aph.gov.au/Parliamentary_Business/Committees/Senate/Finance_and_Public_Administration/Completed_inquiries/2004-07/gallipoli/report/c01 and "Building the ANZAC Commemorative Site" at www.ANZACsite.gov.au/3building/concept.html.

20 P. Malone, "Turkey plans to spend $100m on Dardanelles work: President," *The Canberra Times*, (13 May 2005), 9. See also N. Khadem and F. Walker, "Graves, beach lost for wider road to Gallipoli," *Sunday Age*, (6 March 2005), 7.

21 www.aph.gov.au/Parliamentary_Business/Committees/Senate/Finance_and_Public_Administration/Completed_inquiries/2004-07/gallipoli/report/c04.

22 The catalog, *Legend of Gallipoli Simulation and Information Center* (Ankara, 2012), explains the destruction of the Kabatepe museum building as follows: "It was closed in December, 2007, to be redesigned," 41. See www3.milliparklar.gov.tr/kitap/55/mobile/index.html.

23 Ibid., 27.

24 The official name of the administrative body controlling the Gallipoli peninsula has changed several times in the past five decades. Until 2014 jurisdiction over the region was the responsibility of what is today the Turkish Forest Ministry and Water Affairs and the General Directorate of Nature Protection and National Parks. That ministry was particularly active in the construction of new visitor facilities and oversaw a steady increase in commemorative events and building projects on the peninsula, particularly after the 2002 parliamentary victory of the Justice and Development Party (Adalet ve Kalkınma Partisi, or AKP).

25 OPET has financed several corporate responsibility projects since 2006 as part of their Respect for History Project (Tarihe Saygı Projesi); most recently they have been sponsors of the Alçıtepe mosque complex and the Akbaş Memorial and Monument, discussed in the final chapter. For information in English about the OPET Respect for History Project see the company website www.opet.com.tr/en/about-us.

26 Bahar Bayhan, "Çanakkale Savaşları Gelibolu Tarihi Alan Başkanlığı Kuruluyor," *Arkitera*, 20 June 2014. www.arkitera.com/haber/21751/canakkale-savaslari-gelibolu-tarihi-alan-baskanligi-kuruluyor; İlknur Sudaş, "Gelibolu Yarımadası Tarihi Milli Parkı Yeni Şehitlik Tasarımları Fikir Projesi Yarışması Ön Seçim Sonuçları," *Arkitera*, 4 June 2014. www.arkitera.com/haber/21479/gelibolu-yarimadasi-tarihi-milli-parki-yeni-sehitlik-tasarimlari-fikir-projesi-yarismasi-on-secim-sonuclari.

27 Sarah Midford, *Gallipoli, Anzacs and the Great War* (Melbourne: La Trobe University Press, 2017), 105–19.

28 E. Yeneroğlu and A. Aykaç, "Battlefield Tourism at Gallipoli: The Revival of Collective Memory,the Construction of National Identity and the Making of a Long Distance Tourism Network," *AlmaTourism Journal of Tourism, Culture and Territorial Development* (7), 5, (2016), 69. In "Contemporary Turkish Perceptions of the Gallipoli Campaign," K. Şakul cites on p. 181, n. 4, that one-third of the 1,914 students in a 2011 survey of middle school Turkish students were not aware of the fact that the Ottomans had been defeated in World War One and that 18 March 1915 was the date

of one victory of the Ottomans, not the end of World War One. See also Fatma Çoban, "Orta Öğretim Öğrencilerinin Çanakkale Savaşları ve Öğretiminde Biliş ve Algılama Düzeyleri" (unpublished MA thesis, Çanakkale 18 Mart University, 2011), 94.

29 L. Thys-Şenocak, "Divided Spaces, Contested Pasts: The Gallipoli Peninsula Historical National Park," in *Public Nature: Scenery, History and Park Design*, ed. Ethan Carr, Shaun Eyring, and Richard Guy Wilson (Charlottesville: University of Virginia Press, 2013), 127.

30 Vedica Kant, "Çanakkale's Children: The Politics of Remembering the Gallipoli Campaign in Contemporary Turkey," in *Remembering the First World War*, ed. Bart Ziino (London and New York: Routledge, 2015), 147.

31 Thys-Şenocak, "Divided Spaces", 136.

32 Opening speech given by the rector of Zaim University on 14 March 2015, 4th Gallipoli Symposium, Esenler Municipality, İstanbul. www.izu.edu.tr/Assets/Content/File/_Universite/20150309_4th-gallipoli-sy.pdf.

33 For a recent collection of articles about landscape and conflict archaeology see P. Cornish and N. Saunders, ed., *Bodies in Conflict: Corporeality, Materiality, and Transformation* (London: Routledge, 2014).

34 For studies which bring together landscape, gender and archaeology in interesting ways see S. Nelson, *Handbook of Gender in Archaeology* (AltaMira: Berkeley, 2006); R. Gilchrist, *Gender and Archaeology: Contesting the Past* (London: Routledge, 1999); T. Laqueur, *Making Sex: Body and Gender from the Greeks to Freud* (Cambridge: Harvard University Press, 1990).

35 Some studies doing this are Lucienne Thys-Şenocak and İ. Cerem Cenker, "Moving beyond the Walls: The Oral History of the Ottoman Fortress Villages of Seddülbahir and Kumkale," in *Oral History and Public Memories*, ed. Linda Shopes and Paula Hamilton (Philadelphia: Temple University Press, 2008), 65–86; Lucienne Thys-Şenocak, *Ottoman Women Builders: The Architectural Patronage of Hadice Turhan Sultan* (Ashgate: Aldershot, 2006). See also Zafer Toprak, "Osmanlı Kadınları Çalıştırma Cemiyeti, Kadın Askerler ve Milli Aile," *Tarih ve Toplum*, 51 (March, 1988), 34–8; Yavuz Selim Karakışla, *Women, War and Work in the Ottoman Empire: Society for the Employment of Ottoman Muslim Women (1916–1923)* (İstanbul: Ottoman Bank Archives and Research Centre, 2015).

36 Alan Moorehead's *Gallipoli* (1956) and Nigel Steel and Peter Hart, *Defeat at Gallipoli* (1994) are both good summaries of the Gallipoli campaign, albeit from the perspective of the Allied forces and European sources. Other sources can be found at www.iwm.org.uk/upload/package/2/gallipoli/infoopen.htm and F.R. Van Hartesveldt, *The Dardanelles Campaign, 1915: Historiography and Annotated Bibliography* (Westport, CN: Greenwood Press, 1997). In Turkey the battles fought on the Gallipoli peninsula and on the Asian side of the Dardanelles are referred to as the Çanakkale Wars. For a bibliography of Turkish sources see www.turkeyswar.com/bibliography/; see also E. Erikson, "Strength against Weakness: Ottoman Military Effectiveness at Gallipoli, 1915," *The Journal of Military History* 65, no. 4 (October, 2001), 981–1011; E. Erikson, *Ordered to Die: A History of the Ottoman Army in the First World War* (Westport, CN: Greenwood Press, 2001).

37 Peter Howard, *Heritage: Management, Interpretation, Identity* (London and New York: Continuum, 2003), 21.

38 Tadhg O'Keeffe, "Landscape and Memory: Historiography, Theory, Methodology," in *Heritage, Memory and the Politics of Identity: New Perspectives on the Cultural Landscape*, ed. Niamh Moore and Yvonne Whelan (Aldershot: Ashgate, 2007), 3–4.

39 Several cases of this are examined in M. Gegner and B. Ziino, ed., *The Heritage of War* (London: Routledge, 2012).

40 Serdar Nizamoğlu, presented at the Third Strategies and Activities Conference for the Çanakkale Wars and Gallipoli Historic Region Plans and Site Management Plan, Kolin Hotel, Çanakkale, 23 February 2017.

41 This was pronounced in the Turkish Official Gazette (Resmi Gazete) 29044 and 6546. See http://teftis.kulturturizm.gov.tr/TR,102525/canakkale-savaslari-gelibolu-tarihi-alan-baskanligi-kur-.html which outlines the full list in Turkish of the responsibilities of this directorate. For an assessment of the impact of this administrative change see C. Yürek, "Kentsel Korumada Yönetsel Boyut: Çanakkale Savaşları Gelibolu Yarımadası Tarihi Alan Başkanlığı Örneği," *Akademik Perspektif*, (3) (28 July 2016). http://akademikperspektif.com/2016/07/28/kentsel-korumada-yonetsel-boyut-canakkale-savaslari-gelibolu-yarimadasi-tarihi-alan-baskanligi-ornegi/.

42 UNESCO, Convention on the Protection of the Underwater Cultural Heritage, 2001. www.unesco.org/new/en/culture/themes/underwater-cultural-heritage/2001-convention/.

43 For the list of all criteria possible for nomination to the World Heritage List see http://whc.unesco.org/en/criteria.

44 Ibid.

45 http://whc.unesco.org/en/tentativelists/5911/ as a comparison of the different nomination strategies used for sites in Turkey. The two sites in Turkey which did move from the tentative list to the final list of World Heritage Sites in 2014 were the city of Bursa/Cumalıkızık and Pergamon, both which listed five of the six criteria available to cultural sites for inscription.

46 Jane Lyndon, "Young and Free: The Australian Past in a Global Future," in *Cosmopolitan Archaeologies*, ed. Lynn Meskell (Durham, NC and London: Duke University Press, 2009), 35–9.

47 Ibid.

48 Lynn Meskell, "Cosmopolitan Heritage Ethics," in *Cosmopolitan Archaeologies*, ed. Lynn Meskell (Durham, NC and London: Duke University Press, 2009), 14.

49 John McQuilton, "Gallipoli as Contested Commemorative Space," in *Gallipoli: Making History*, ed. Jenny Macleod (London and New York: Frank Cass, 2004), 151.

50 Josie Taylor, "Danna Vale Suggests Australian Replicate of Gallipoli," *AM Monday* (Melbourne: Australia Broadcasting Corporation, 2005); www.abc.net.au/am/content/2005/s1484587.htm.

51 "Treaty of Lausanne," 1923, The World War I Document Archive; https://wwi.lib.byu.edu/index.php/Treaty_of_Lausanne.

52 Discussions regarding the construction of roads on the peninsula can be found at "*Matters Relating to the Gallipoli Peninsula*," The Senate, Finance and Public Administration References Committee, Commonwealth of Australia, www.aph.gov.au/SEnate/committee/fapa_ctte/gallipoli/report/index.htm. For the 6 June 2011 agreements between the Turkish and Australian governments concerning the prevention of erosion at Arıburnu and improvement of the ANZAC cove area see http://minister.dva.gov.au/media_releases/2011/jun/va039.pdf.

53 Rosie Ayliffe, Marc Dubin, and John Gawthrop, *Turkey: The Rough Guide* (London: 1991)

54 For a longer discussion of this see L. Thys-Şenocak, "Divided Spaces, Contested Pasts: The Gallipoli Peninsula Historical National Park," in *Public Nature: Scenery, History and Park Design*, ed. Ethan Carr, Shaun Eyring, and Richard Guy Wilson (Charlottesville: University of Virginia Press, 2013), 127–40.

55 Some of these studies are by George Cassar, *The French and the Dardanelles: A Study in Failure in the Conduct of War* (London: George Allen and Unwin, 1971); Philip Lecane, *Beneath a Turkish Sky: The Royal Dublin Fusiliers and the Assault on Gallipoli* (Gloucestershire: The History Press, 2015); Jenny Macleod, "Britain and Ireland: Gallipoli Day or Anzac Day?" *Gallipoli* (Oxford: Oxford University Press, 2015), 126–53.

56 For the issue of the management and museum display of archaeological human remains in Turkey, see L. Thys-Şenocak and E. Doğan, "Archaeology, Museums and Tourism on the Gallipoli Peninsula: Issues of Human Remains, Ordnance, and Local Decision-Making [Gelibolu Yarımadası'nda Arkeoloji, Müzeler ve Turizm: İnsan Kalıntıları, Mühimmat ve Yerel Karar Alma]," in *Heritage in Context II*, (2018), 289-313.

57 Law on the Conservation of Cultural and Natural Property (2863) can be found at www.kulturvarliklari.gov.tr/TR,43249/law-on-the-conservation-of-cultural-and-natural-propert-.html. For information related to cultural materials from the time of the National Independence War and the Foundation of the Republic of Turkey see *Turkish Constitution, Chapter 3*: "Movable cultural and natural property to be protected: Article 23. The following shall be movable cultural and natural property to be protected: b) Due to their importance for national history, documents and goods of historic value relating to the National Independence War period and the Foundation of the Republic of Turkey, personal belongings, documents, books, correspondences and similar movables of Mustafa Kemal Atatürk.

58 UNESCO "Cultural Landscapes" (2017) http://whc.unesco.org/en/culturallandscape/.

59 Turkish Ministry of Culture and Tourism, ÇATAB 2016–2020 Stratejik Planı can be accessed at http://catab.kulturturizm.gov.tr/Eklenti/50049,2016–2020-stratejik-planpdf.pdf?0. On 24 February 2017 at the third planning conference organized by ÇATAB for the heritage management of the Gallipoli peninsula, İclal Dinçer, the director of Turkey ICOMOS from 2014–17, proposed officially that the Gallipoli peninsula should be categorized as a cultural landscape. Turkey has only recently had World Heritage sites inscribed which were designated as cultural landscapes, with Pergamon in 2014 and Diyarbakır in 2015. http://whc.unesco.org/en/statesparties/tr.

60 D. Cosgrove, "Introduction to Social Formation and Symbolic Landscape," in *Landscape Theory*, ed. R. DeLue and J. Elkins (New York and London: Routledge, 2008), 20. See also Denis E. Cosgrove, *Social Formation and Symbolic Landscape* (Madison: University of Wisconsin Press, 1984).

61 *Heritage, Memory and the Politics of Identity: New Perspectives on the Cultural Landscape*, ed. Yvonne Whelan and Niamh Moore (New York: Routledge, 2016), x.

62 Henri Lefebvre, *The Production of Space*, trans. Donald Nicholson-Smith (Oxford and Cambridge, MA: Basil Blackwell, 1991), 286.

63 Peter Howard, *Introduction of Landscapes* (Burlington: Ashgate, 2011), 1.

64 Pierre Nora, "Between Memory and History: Les Lieux de Mémoire," *Representations 26* (1989), 7–24.

65 Jay Winter, *Sites of Memory, Sites of Mourning* (Cambridge: Cambridge University Press, 1995), 1.

66 R. Nuwer, "There's no such thing as truly pristine nature anymore." (8 February 2016), BBC; www.bbc.com/future/story/20160208-theres-no-such-thing-as-truly-pristine-nature-anymore.

67 Simon Schama, *Landscape and Memory* (Toronto: Random House, 1995), 61.

68 Peter Howard, *Introduction of Landscapes* (Burlington: Ashgate, 2011), 1.

69 It is also crucial when using the category of cultural landscapes, precisely because of its wide appeal for many different discplines, that readers are not alienated by specialized terms and jargon which do not cross disciplines easily. For that reason I have purposely limited the use of specialized heritage terminology in this book, so that it makes sense beyond the heritage field.

70 (UNESCO 2017); http://whc.unesco.org/en/culturallandscape/#1.

71 David Lowenthal, "Past Time, Present Place: Landscape and Memory," *Geographical Review* 65 no. 1 (1975), 36.

72 (UNESCO 2017) http://whc.unesco.org/en/tentativelists/5911/.

73 Alexander Scott, "Russell Crowe's Water Diviner Tries to Question History, but Misses the Mark," *The Conversation*, 7 April 2015, http://theconversation.com/russell-crowes-water-diviner-tries-to-question-history-but-misses-the-mark-39519.

74 Brad West, "Enchanting Pasts: The Role of International Civil Religious Pilgrimage in Reimagining National Collective Memory," *Sociological Theory* 26 no. 3 (2008), 266.

75 Kenneth Stanley Inglis and Jan Brazier, *Sacred Places: War Memorials in the Australian Landscape* (Melbourne: The Miegunyah Press, 2008), 383–4.

76 www.smithsonianmag.com/arts-culture/russell-crowe-takes-new-look-at-battle-gallipoli-180954962/.

77 Azer B. Kemaloğlu, "Anniversaries and Production of Fiction: Gallipoli," in *War, Memory and Commemoration*, ed. Brad West (New York and London: Routledge, 2017), 198.

78 Jonathan Alford, ed., *Greece and Turkey: Adversity in Alliance* (Aldershot: Gower, 1984), 91.

79 *Gelibolu Peninsula Historical National Park* (Turkish Ministry of Environment and Forestry General Directorate of Nature Protection and National Parks, Ankara: 2005), 44.

80 The Çanakkale Destanı has been translated into English by the Ministy of Forestry and Water Affairs as the Legend of Gallipoli. This is the name which will be used throughout this book for the Legend of Gallipoli Simulation and Information Center.

81 www.milliyet.com.tr/erdogan-canakkale-zaferinin-101-/siyaset/detay/2211796/default.htm.

82 In Turkish this is "Şehitler ölmez, vatan bölünmez."

83 For a recent survey of the ANZAC sector of the Gallipoli peninsula, see *Anzac Battlefield: A Gallipoli Landscape of War and Memory*, ed. A. Sagona, et al.; Philip Lecane's *Beneath a Turkish Sky: The Royal Dublin Fusiliers and the Assault on Gallipoli*, (Gloucestershire: History Press, 2015) is an exception to the ANZAC-focused studies of the peninsula's events and geographies as he presents an account of the experiences of the Allied Forces in the Cape Helles region and particularly the role of the Irish troops.

84 David Lowenthal, "Past Time, Present Place: Landscape and Memory," *Geographical Review 65* no. 1 (1975), 5.

2 Cultural landscapes of the Gallipoli peninsula

From mapping to metaphor

Among the Gallipoli peninsula's many layers of cultural landscapes, the one which existed between 1914 and 1916 can be considered as a specific type of heritage: the battlefield terrain. The 2014 Vimy Declaration for the Conservation of Battlefield Terrain outlines the definition, preservation, and management of such landscapes:

> Battlefields are key sites in a continuous educational process. These are places where successive generations revise or expand their cultural memory through interaction with the artifacts and landscape of the past. Yet battlefields are not a single, sealed terrain isolated in a given moment of time. Rather, they are multi-vocal landscapes of accretion, stratified by overlapping layers of social, economic and political history, and almost infinitely connected in myriad ways with other places: where soldiers came from, for example. They are also invariably politicized, dynamic and open to constant negotiation. Battlefields are often contested landscapes, and can become the focus for new conflict.[1]

In the Vimy Declaration, the fluid boundaries of time and place that characterize the historic battlefield are acknowledged. Returning this type of landscape to its "authentic" state, to a time when a famed battle was fought, is not considered a viable or even advantageous undertaking. Rather, the document continues, "One can only allude to the original conditions, not recreate them."[2] Landscapes, like historical monuments, are continually changing; they have had lives and several identities before, during, and after the specific events that transformed that landscape into a battlefield. Like other battlefield landscapes, the Gallipoli peninsula is "not a single, sealed terrain isolated in a given moment of time"; it exists as a prime example of "multi-vocal landscapes of accretion, stratified by overlapping layers of social, economic and political history."[3]

As a cultural landscape, the Gallipoli peninsula has been shaped by centuries of continual interaction between humans and the terrain. The battlefield landscape of Gallipoli is one phase among many of the relationships that humans have had with this region. These have ranged from mapping to physical manipulation of its topographic features to de- and re-forestation, to ascribing it with poetic metaphors of embodiment.

Recording the landscape: early mapping projects on the peninsula

Prior to the actual landing of Allied troops on the beaches of the Gallipoli peninsula on 25 April 1915, an extensive amount of data about the region's flora, fauna, and topography had been gathered by European diplomats, spies, travelers, and Ottoman cartographers. While cartographic documentation of the Dardanelles had been produced many centuries earlier, the interest in recording the topography of the peninsula increased in the late nineteenth and early twentieth centuries, with the products of several of these reconnaissance and mapping missions ending up in the military and naval archival repositories of Turkey, Britain, Australia, Germany, France, and Russia.[4]

French cartographers, hydrographers, naval and military envoys, and ambassadors to the Ottoman Empire in the eighteenth and nineteenth centuries (such as Barbié du Bocage, M. le chevalier de Clairac, Choiseul-Gouffier, and Lafitte-Clavé) all had different incentives for documenting this part of the Ottoman Empire, and each left detailed descriptions of the landscapes of many of the regions visited, including the Gallipoli peninsula and the Dardanelles.[5] Many had antiquarian as well as military interests and noted in passing how ancient ruins from earlier civilizations on the straits were being refashioned into marble cannon balls, or how column shafts from Greco-Roman sites were recycled as spolia to build harbors and seaside fortifications on the Dardanelles.[6] The French had a particularly active role in the late nineteenth-century documentation of Gallipoli's landscape. Topographical maps of the peninsula, such as the *Carte de la Presqu'île de Gallipoli, Levée par les Officiers d'État-Major de l'Armée d'Orient*, were produced by French cartographers during the Crimean War (Figures 2.1(a) and (b)).

Many of the late nineteenth century Ottoman cartographers who mapped the peninsula were trained by the French in Paris, and then in Istanbul where they helped to establish the Military Mapping Department in the Ottoman Empire in 1895.[7] Ottoman interest and production of maps of the Dardanelles region had a long history; among the earliest of these projects in the empire was Piri Reis' famed atlas of sixteenth-century seascapes, the *Kitab-ı Bahriye*[8] (Figure 2.2). Mapping efforts of the Dardanelles by Ottoman engineers intensified after the establishment of the Imperial School for Ottoman Engineers (the *Mühendishane-i Berrî-i Hümâyûn)* in 1794, and the *Mekteb-i Harbiye*, or Military Academy, by Sultan Mahmud II in 1834. Among the products of these schools were several maps of the Dardanelles showing the location of ordnance and the different sets of fortresses along the European and Asian shores.[9]

Well-known for his extensive mapping of the post-WWI Gallipoli peninsula, Mehmet Şevki Pasha was among the first of the Ottoman cartographers trained in Paris from 1890 to 1894, and he later became the geodesy director of the Ottoman Mapping Commission in 1909. As trained by the French, he selected a Bonne projection for his maps whose central meridian was the crescent on the dome of the Haghia Sophia mosque in Constantinople.[10] By May 1915, thanks in

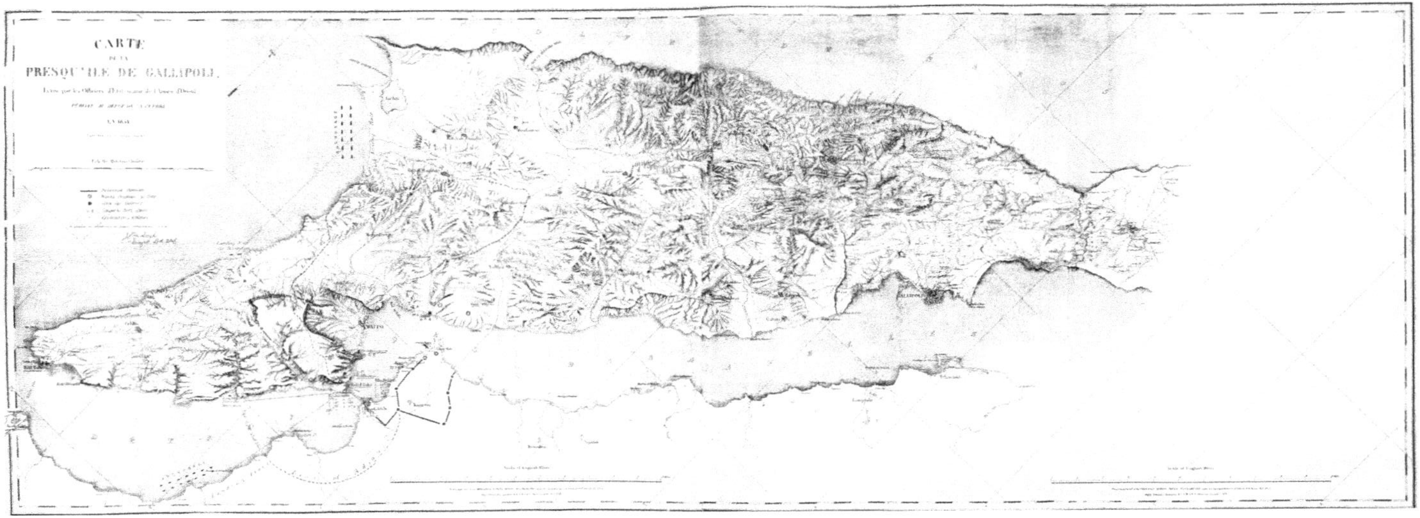

Figure 2.1 'Carte de la Presqu'île de Gallipoli, Levée par les Officiers d'État-major de l'Armée d'Orient' 1854. Overprinted at the Intelligence Branch, Quarter Master General's Department, 1880, to show defensive positions and works proposed in 1880 for armaments, landing places, anchorages.

Credit: The National Archives, Kew, MPH 1/803/2–3

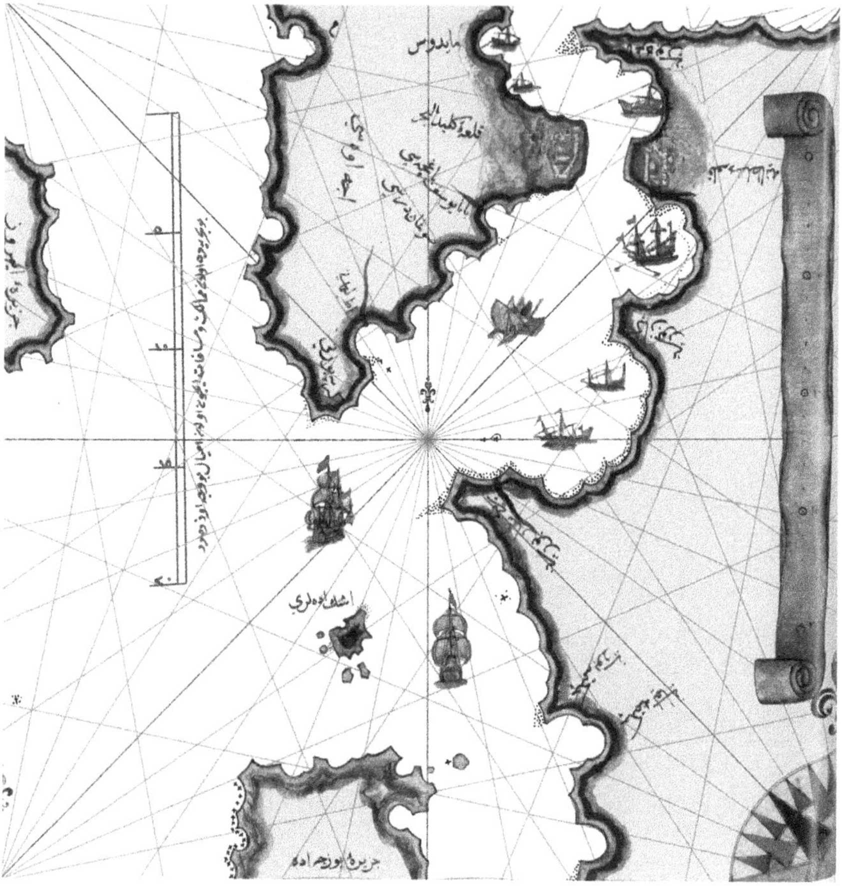

Figure 2.2 Detail of a map of the Dardanelles showing the fifteenth century fortresses of Sultaniye (Çanakkale) and Kilitbahir on the Straits, possibly made by Piri Reis in 1525/26.

Credit: İstanbul University Library, Rare Book Collection, T6605

part to Şevki Pasha, the Ottomans had produced several maps of most sections of the Gallipoli peninsula at differing scales of 1:20,000, 1:25,000, and 1:50,000.[11] A comprehensive set of maps for the entire peninsula completed by the Ottomans in 1916 was among the main resources used by the Ottoman sultan's commanders immediately after World War One to identify both Allied and Ottoman features of the battlefield, and to reconstruct the course of events that had transpired during the battles fought there[12] (Figure 2.3). The maps of Şevki Pasha continue to have a significant impact on the planning of the peninsula. Both the Turkish Ministry of Forest and Water Affairs and ÇATAB have utilized these maps to decide where to place commemorative buildings, tourist routes, symbolic cemeteries, museums, and visitor centers on the peninsula.[13]

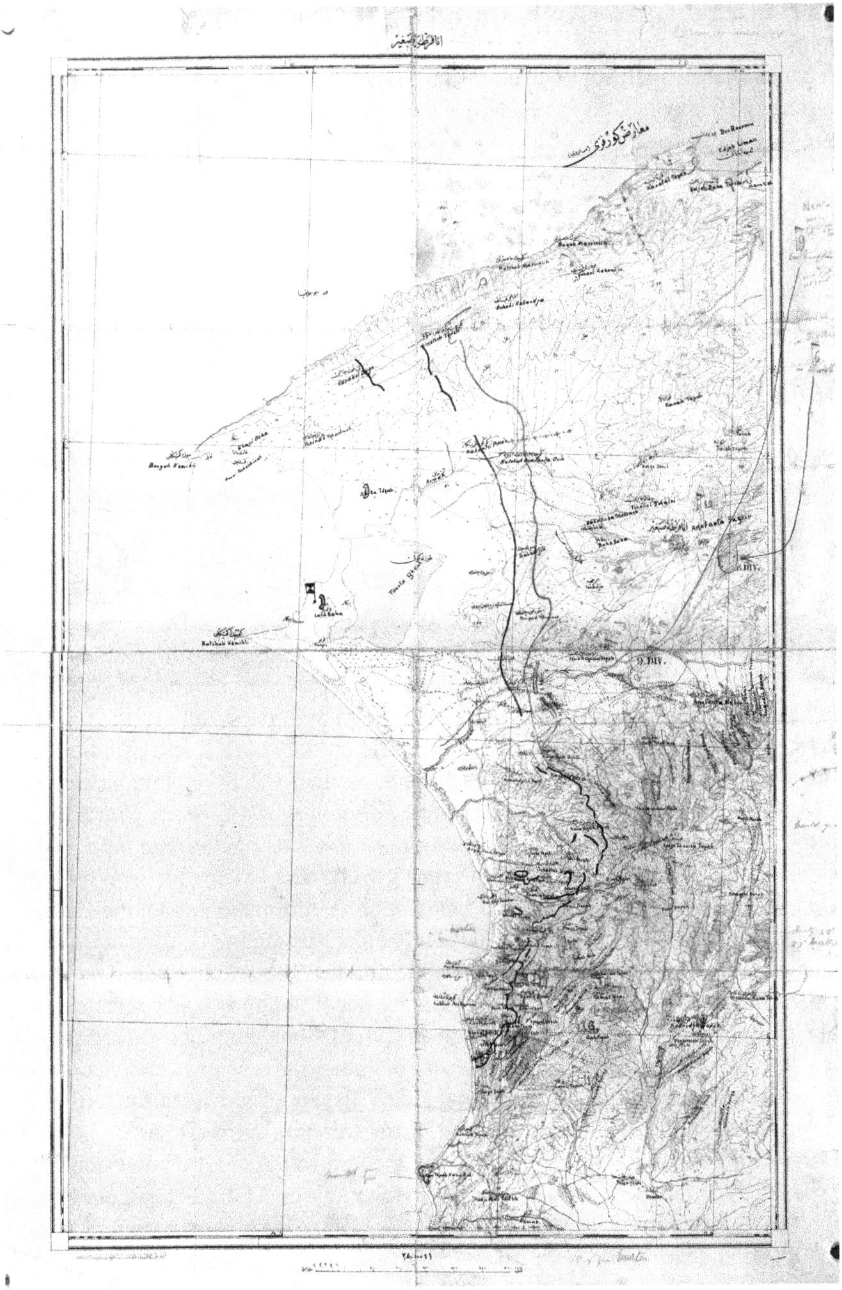

Figure 2.3 Şevki Pasha's map. One of 43 maps showing the Gallipoli peninsula, surveyed by Brigadier General Mehmet Şevki Pasha after the evacuation of Gallipoli in December 1915 and January 1916.

Credit: Australian War Memorial RC02688–1

In addition to the French, the British were also actively involved in the mapping of the Dardanelles and the Gallipoli peninsula as part of their military reconnaissance in the Ottoman Empire. British naval hydrographers Thomas Spratt and Thomas Graves conducted surveys of the peninsula and the region of the Dardanelles in the mid-nineteenth century, the results of which were compiled to produce the Admiralty Chart of 1871, referred to as "*The Dardanelles (Ancient Hellespont) From the West Entrance to Cape Nagara.*"[14] In 1876 C.E. Grover, working for British intelligence, measured several sections of the Gallipoli landscape including the future landing area of ANZAC cove.[15] Thomas English, of the Geological Society of London, conducted an on-site geological survey of the Dardanelles in 1904, and another in 1912. Two comprehensive pre-WWI reports on the topography of the Gallipoli peninsula by the British came out of the Naval Intelligence Department in 1909.[16] Dr. Walter Leaf's book *Troy, A Study in Homeric Geography* was published in 1912, along with photographs, including several of the fortresses on the Gallipoli peninsula. This research, and its potential use for planning battle strategies in World War One, went largely unnoticed until after the war.[17]

There was a long-established tradition of coupling scholarship about the Dardanelles with military reconnaissance, and the mapping missions just prior to World War One were no exception. Most of the information gathered at this time by the British and French focused on the state of Ottoman fortifications at the entrance to the Dardanelles and the potential areas for landing in a military invasion. Their interest at this time was clearly not in the beauty of the Gallipoli landscape but motivated primarily by strategic military objectives, as indicated in the unabashed titles of some of these reports such as the 1876 pamphlet entitled "*The Dardanelles for England: The True Solution of the Eastern Question,*" or the 1880 paper "*Seizure of the Dardanelles as a means of coercing the Porte.*"[18]

The Germans, with whom the Ottoman government would side in World War One, had also sought and recorded topographical information about the Gallipoli pre-war landscape more than a decade before the outbreak of battles there. Heinrich Kiepert was largely responsible for the 1880 *Karte von Kleinasien*, which included maps of the Dardanelles. Kaiser Wilhelm II was concerned with keeping good relations with the Ottoman Sultan Abdülhamid II to facilitate the expansion and security of the Berlin to Baghdad railway line, and maps were essential for this. Like France, Germany also played a key role at the turn of the nineteenth century in helping the Ottomans to shore up their defenses in the Dardanelles and had been involved in the modernization of Ottoman troops prior to and during the Balkan Wars. General Helmuth von Moltke had spent time in the Gallipoli region in 1836; Colmar von der Goltz led a German mission there in 1883 and again in 1908.

The Germans produced a wealth of information about the topography of the Gallipoli peninsula just prior to the outbreak of World War One and immediately after the Second Balkan War in 1913, including a book by German Lieutenant Georg Hans Rohde, *Die Operationen an den Dardanellen im Balkankriege 1912–13*, which included maps, photographs, and other vital information of potential use to military operations. General Liman von Sanders, the German General of Cavalry in charge of the Dardanelles region during the Gallipoli battles, began his service to the sultan as Chief of the Fifth Army for the Ottomans early in 1915[19] (Figure 2.4).

While these military reports of the Gallipoli topography focus on its terrain and features such as the height of mountains, length of ridges or beaches, and possible water sources for land troops, occasionally there are other aspects of the landscape mentioned, including what was growing wild on the peninsula and the type of agriculture being practiced there prior to the war. In the 1909 War Office Report, the presence of low and thick scrub in the area around Suvla Bay is frequently mentioned as a possible impediment to the movement of troops. Other observers described "dwarf ilex . . . with limbs frequently as thick as a man's arm, and with

Figure 2.4 Liman von Sanders photographed by Bain.

Credit: Library of Congress, Public Domain, https://commons.wikimedia.org/w/index.php?curid=29840477

foliage through which it was impossible to force one's way."[20] But several areas of the peninsula were quite suitable for farming, and described as "covered with cultivation and dotted with large trees, oak and walnut."[21] The area around the Azmak stream was "greener and supported several tree-ringed farmsteads."[22] Another report, prepared by the General Staff of the War Office just a decade before the war began, specified that the resources of the peninsula included "corn, but cotton is also grown. Timber is scarce but about 12 miles below Gallipoli, at Ungardere and Yalova, are small fir woods."[23]

Deforestation of the peninsula in the Ottoman era

Considering how important environmental history has been for the reinterpretation of the Mediterranean past, it is surprising how little research exists on the environmental history of the late Ottoman Empire or specifically the Gallipoli landscape.[24] Information about the flora and fauna from earlier eras of the peninsula does appear in the writings of Classical Greek and Roman writers. As Isaac has noted, "the Chersonese was more than a piece of land dominating important straits. It was a flourishing region inhabited by a remarkable variety of Greek communities, ranging from small fortified villages to substantial cities."[25] Studies about the natural environment of the Chersonese have been conducted using ancient texts such as the *Iliad* and these have revealed that the landscape was once populated by various species of oak, such as the *Quercis robor*.[26] Other trees such as the elm, tamarisk, willow, and pine were also mentioned by Homer as present in the region.[27] In addition, ancient sources noted that there was a large difference between the northern and southern lands of the Gallipoli peninsula and that its northern section was a fertile place for agriculture, especially the fields and crops planted in the red loam.[28] Xenophon, writing in the fifth–fourth century BCE, noted that the lands bordering the Dardanelles were prosperous and that some of the settlements on the peninsula, of which there were around twelve, were occasionally named after the product grown there[29] (Figure 2.5).

Elaious (Eléonte), founded in the late seventh century BCE on the southern part of the peninsula at Morto Bay, and now under the Çanakkale Martyrs Memorial, was named for its olive trees; Krithea (Kirte/Alçıtepe) was known for barley; and Athens was a major exporter of wheat from the Hellespont region.[30] Other Classical-era writers report there were several suitable harbors; corn, fruit trees, and cattle were also to be found. The fertile lands of the region attracted the Thracians to the north, prompting Miltiades I in the sixth century BCE to construct a wall which stretched across the northern borders of the Gallipoli peninsula to protect its cities and farmlands from these and other invaders. Athenians settled on the Gallipoli peninsula by the middle of the sixth century BCE and forced the Persians off the land; thereafter it became part of the famed Delian League and was controlled by the powerful city-state on the Greek mainland.[31]

In the sixth century CE the Byzantine Emperor Justinian rebuilt in stone the earlier sixth century BCE wall, which was most likely an earthen fortification. Two fortresses on either side of the straits, at the cities of Abydos and Sestos,

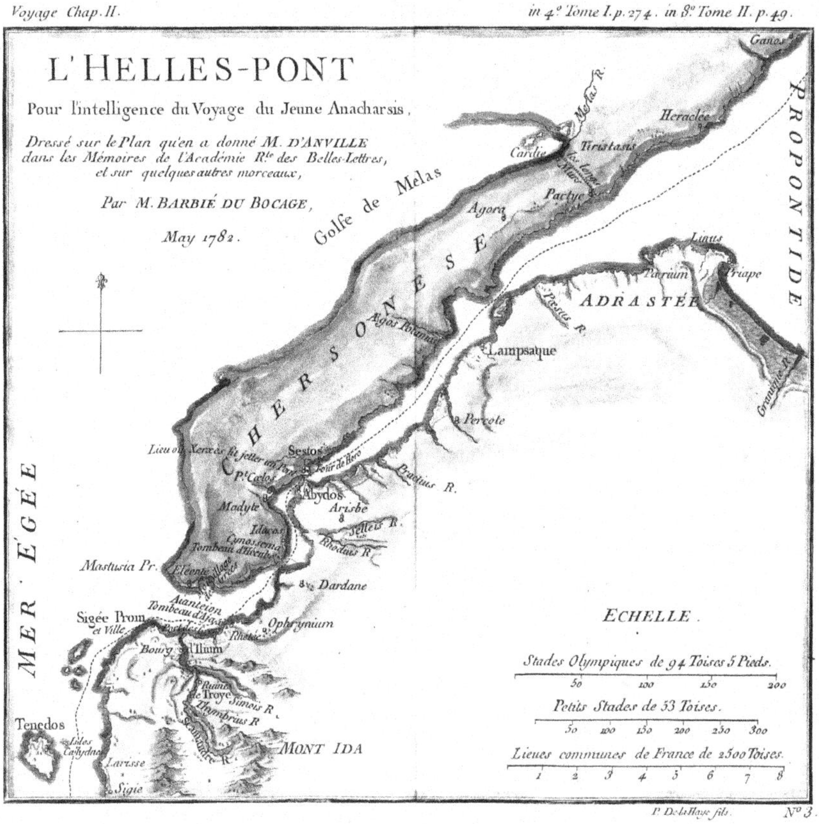

Figure 2.5 Map of the Ancient Hellespont by Barbié de Bocage, May 1782. Antique copper plate engraving indicating the sites of Elaious (Eléonte), Sestos, Abydos, and other Classical and Roman sites on the straits.

Credit: Archive of L. Thys-Şenocak

were also constructed at this time to collect customs tax and protect the interior lands of the Byzantine Empire and its capital Constantinople from enemies. In addition to continuing the earlier agricultural practices of their Greek and Roman ancestors, the Byzantines practiced viticulture on the peninsula, as evidenced by recent archaeological discoveries of wine-making facilities in Çamburnu and Maydos (today's Eceabat).[32]

The Çanakkale straits were of particular importance in the expansion of the Ottomans into Europe. The capture of the peninsula and the Byzantine fortress of Çimpe in 1352 by Orhan Bey and his son Süleyman Pasha marked a turning point in the Ottoman invasion of Thrace, giving the Ottomans their first foothold in Europe. From these early Ottoman victories in Thrace through the seventeenth century, the protection of the straits and the Gallipoli peninsula was of utmost importance for both the strategic military and trade advantages that the possession of this geography gave the Ottomans in the eastern Mediterranean, and the natural resources that the lands in these regions held.[33] The need to defend the straits also led to the depletion of some of its resources such as timber.

Contributing to the deforestation of the peninsula were the fortifications and other monuments that were built by members of the imperial Ottoman family, as the protection of the straits became increasingly significant for the expanding empire. Two fifteenth-century fortresses on the Gallipoli peninsula lent Mehmed II's imperial Ottoman stamp to the region. This formidable pair of fortresses served to protect the gateway to the capital of the empire; as massive construction projects, Kilitbahir and Kale-i Sultaniye, or Çimenlik, the fortress Mehmed II constructed in today's Çanakkale, tied this sultan's imperial legacy to the earlier heroes—real and mythical—of antiquity who had battled against intruders in the region of the Dardanelles. Mehmet II's Greek chronicler Kritovoulos noted that when the two fortresses of Kilitbahir and Kale-i Sultaniye were completed, they could "separate the upper sea (Black Sea) from the lower (Aegean), and to prevent completely and make it entirely impossible the navigation up or down for any who might wish it, except for those whom he himself was willing to allow pass . . ."[34] Over two centuries later the depiction in 1700 of Kilitbahir, Mehmed II's fortress on the European shore by the royal French engineer Nicolas Berquin shows in great detail the architecture and much of the ordnance of the fortress, including drawings which demonstrate how the cannons there were loaded. This drawing was among several made by French military engineers as part of their extensive military reconnaissance of the geography and defensibility of the Gallipoli peninsula from the late seventeenth through the nineteenth centuries[35] (Figure 2.6).

More timber resources from the Gallipoli peninsula and its surrounding forests were consumed two centuries later when two additional fortifications, Kumkale and Seddülbahir, built by Turhan Sultan, the queen mother (or *valide)* of Ottoman sultan Mehmed IV, were needed to supplement the defense capabilities of the pair of earlier fortifications at Kilitbahir and Kale-i Sultaniye.[36] While the Ottoman chroniclers and Turhan Sultan's foundation charter (*vakfiye*) sang the praise of the new fortresses at the entrance to the Dardanelles and extolled the munificence of their patron, a building project on such a massive scale took a toll on the environment.[37] As is apparent from the sparse vegetation depicted in Nicolas de Fer's engraving of the Dardanelles in 1683 the building and repairs made to Ottoman fortresses also accelerated the deforestation of the Gallipoli peninsula and the surrounding regions (Figure 2.7). Along with the timber harvested for the production of ships for the Ottoman navy, forests were sacrificed for the wooden structures within the fortresses, stabilizing beams needed for their thick defense walls, and scaffolding, firewood, and charcoal for imperial building projects both on the peninsula and in the Ottoman capital. All contributed to the deforestation of the lands surrounding the Dardanelles.[38] The seasoned seventeenth century traveler and Ottoman chronicler Evliya Çelebi described in the mid-seventeenth century the forests in this western region of present-day Turkey, from Izmit to Sakarya, as an "*ağaç denizi*" or a "sea of trees."[39] A half century later Aaron Hill,

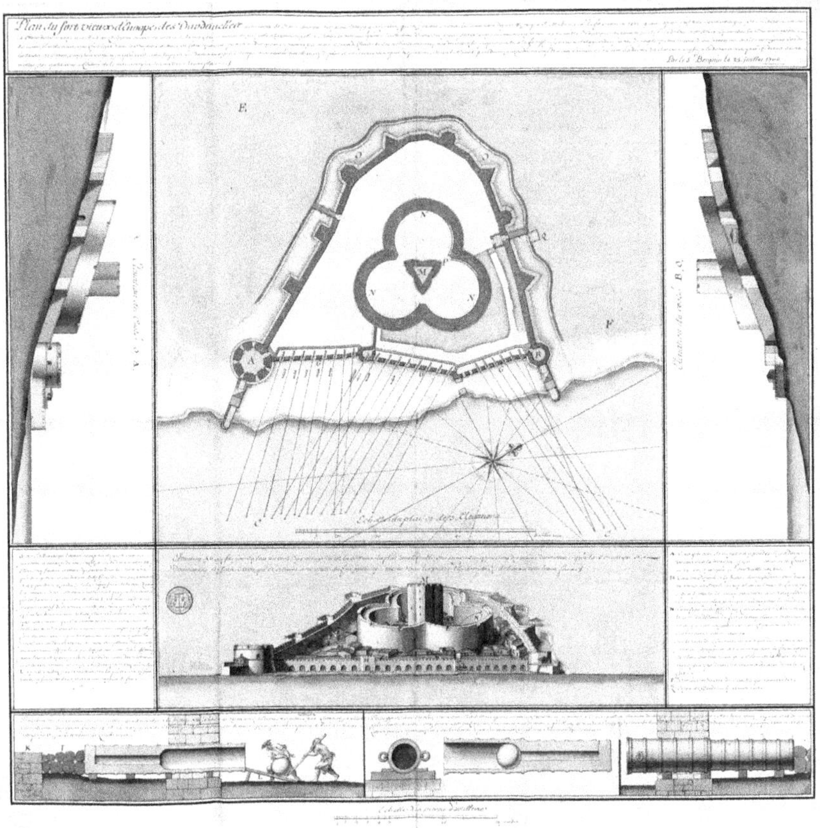

Figure 2.6 Drawing from 1700 by the French military engineer Berquin of the Kilitbahir fortress, which was built by the Ottoman Sultan Mehmed II in the fifteenth century, and detail (below) of how the cannons there were used.

Credit: BnF 1700 70 × 72 cm BnF cartes et plans GE SH 18 PF DIV 5 P2

in his early eighteenth century travel account *"A Full and Just Account of the Present State of the Ottoman Empire in all its Branches With the Government and Policy, Religion, Customs, and Way of Living of the Turk in General,"* noted that the forests on both the Black Sea and the shores of the Gallipoli peninsula were being regularly harvested for construction projects in Istanbul: "round the shores whose ragged Cliffs encompass the Euxine or Black Sea, stand endless Number of Tall Woods, whose stately Timber fell'd for use supply the city thro' the Thracian Bosphorus, with the finest Wood required for Buildings."[40]

Prior to the nineteenth century the majority of Ottoman forests surrounding the Black and Caspian and Mediterranean seas, including the forests of the Taurus mountains, were considered to be *miri* lands, i.e., the property of the sultan; technically, no harvesting of the wood by others was allowed without permission.

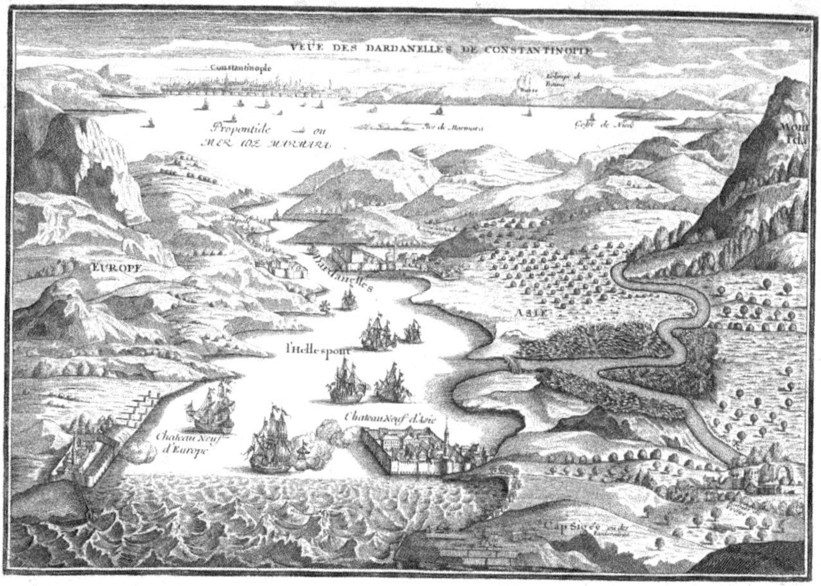

Figure 2.7 Veüe des Dardanelles de Constantinople, Nicolas de Fer (from Grelot's *Relation nouvelle de Constantinople*, 1683), showing the two sets of Ottoman fortresses from the fifteenth and seventeenth centuries and a deforested landscape.

Credit: Archive of L. Thys-Şenocak

The actual management and protection of forests was done locally, however, so ultimately there was no effective policy of protection for forests in the Ottoman Empire in the pre-industrial era. The empire's forests, like Egypt's fields of grain, were considered by the sultan and his central administration to be its bread basket and were depleted accordingly. Ottoman forests, therefore, stood at this time as an immense repository of wood to be consumed by the construction of fortifications and other imperial building projects, such as the shipbuilding industry of the sultan's navy. The timber needs of the arsenal, with its 300 to 400 galleys, contributed to the serious deforestation in the pre-industrial Ottoman era, with the coastal forests being particularly hard-hit.[41] In addition to these naval projects, once harvested, the forests of Anatolia and Thrace reappeared as waterwheels, dams, and supports for embankments along the Nile, wooden storage casks for the water needed by Ottoman sailors crossing the Mediterranean, or as a trading commodity to be sold for other luxuries such as the pepper and spices needed in the palace kitchens (Figure 2.8).[42]

The administrative offices of the Ottoman Imperial Dockyards were responsible for overseeing the utilization of the empire's forests, granting some as *Cibal-i Mübaha* (mountains legally available for common use/common forests) for the use of the peasants who needed firewood to burn and timber to build.[43] Archival

Figure 2.8 Repairs made on the shores of the Dardanelles to ships like the San Pascual of the Spanish fleet depleted nearby forests, from José Moreno, *Viage a Constantinople*, 1784 (Madrid, 1790).

Credit: Aikaterini Laskaridis Foundation

documents attest to an ongoing battle with locals over the use of forest resources. An order sent from Sultan Süleyman I to the *kadı* (judge) of Gallipoli in September 1565 requests that the local administration ensure that the forests belonging to the sultan (*koru-yı hümayun*) were not to be harvested by others so that an ample supply of timber could be accessed for imperial shipbuilding needs.[44] Those who violated these restrictions faced severe punishment and confiscation of their goods, as was the case with a female merchant whose shipment of 70,000 staves for wine barrels was seized en route to France.[45] Locals in Thrace who ignored these imperial decrees and felled trees or even cut the upper branches of trees in the royal forests were sent to labor in the sultan's galleys.[46]

However, the depletion of forests on the Gallipoli peninsula, legally and illegally, continued unabated until the Ottomans began to approach forestry management more scientifically. The modernizing reforms of the Tanzimat era in the nineteenth century witnessed an increasing professionalization of the bureaucratic apparatus related to forest management. In 1839–40 the Forest Ministry was established by the Ottoman government, and initially was tied to the Ministry of Trade; by 1869 it was operating under the auspices of the Ministry of Finance.[47] Sultan Mahmud II started the practice of bringing in foreign specialists to advise on the utilization of the resources of the empire's forests; the earliest of these was Gustave de Pauliny, an Austrian mining expert who became the Director-General of Mines in 1836. He conducted surveys of the forests surrounding important

Ottoman mines, particularly in the regions of the empire near present-day Tokat and Diyarbakır, as these studies were needed for determining the costs of the production of charcoal used in the mining industry.[48] Between 1857 and 1878 the Ottomans received much of their forest management guidance from the French who had been invited by the sultan to share their expertise on the European practices of silviculture and *aménagement*.[49] Many foreign observers had noted, including George Perkins Marsh, an early environmentalist from the United States who had spent part of his overseas career in the Ottoman Empire as a diplomat, that the practices in the Ottoman Empire and in many parts of the Mediterranean were unsustainable and had left the Mediterranean forests as "... deserted by civilized man and surrendered to hopeless desolation."[50]

Under advisors such as Louis François Tassy, the founder of the first Ottoman Forestry School and author of its first textbook, *Études sur l'aménagement des Forêts*, surveys of the forests in Ottoman lands began with the first eight graduates from the school in 1861.[51] Forests along the coasts (and whose wood could be more easily harvested and transported abroad) were among the first to be accounted for; the stretch of forest lands near Çanakkale and the coastal forests between the Dardanelles and Tekirdağ, including the Gallipoli peninsula, were among the earliest to be surveyed and designated as profit-making regions.[52] Reports to the survey commission from the governor of Midilli (Lesbos) on the deforestation problem of that island give some indication of the kind of threat that forests in the northern Aegean region of the empire, including the Gallipoli peninsula, must have been facing. Among these were fires started by villagers who were interested in usurping land for agricultural or grazing purposes.[53]

Motivated by an interest both within and outside of the Ottoman Empire's boundaries to better utilize the resources of the empire's forests, the Forest Regulation of 1870 was proclaimed on 14 January 1870. It drew from the legislation related to forests in the Land Code of 1858 and a rather loose set of forest legislation proclaimed in 1861, as well as the expertise of the French forestry experts, to create a more cohesive set of policies which "defined the prohibitions in forests, the rights and 'privileges' of local people, of the merchants and contractors, the supply of (the) Imperial Shipyard and Artillery, browsing in forests, and finally, the penalties for forest offenses."[54] The basic tenets of this 1870 legislation remained in effect, with some minor changes, until 1937. Bringing together these different Ottoman laws was not an easy task for the Ottomans, and the results were sometimes confusing for their European advisors. Hyde Clarke, a British member of the Imperial Land Commission who had been tasked with conducting land surveys in the Ottoman Empire, remarked that according to some of the stipulations in the Land Code of 1858, unlike the law of land ownership in Europe, occasionally the ownership of trees preceded the ownership of the land. "The trees," he concluded, "may have given ownership in the land, rather than the land giving ownership in the trees."[55]

Promulgating the Forest Regulation of 1870 was intended as a step to better utilize the resources of the Ottoman empire's lands, and it reflected the increasing bureaucratization of forestry practices that was common throughout Europe in the late nineteenth century. This era marked an important pre-WWI stage in the documentation, mapping, and assessment of forest resources, a shift from publicly-used to state-controlled resources, and the establishment of new state institutions to oversee the maintenance and exploitation of the empire's forests including those in the Gallipoli region.[56] The changing status of the Forest Ministry in the larger Ottoman administration is indicative of its importance to the government for generating revenue, and reveals there was a growing interest in capitalizing on natural resources that were becoming scarce. From 1869 to 1908, the administrative group appointed to manage the forests in the empire shifted several times; initially forest management was under the direction of the Finance Ministry, then the Ministry of Forests and Mines, the Ministry of Trade and Agriculture, and finally the Ministry of Forests, Mines and Agriculture. The period from 1908 through 1920 saw a relatively long stretch of administrative stability as the management of the empire's forests came under the control of the Ministry of Trade and Agriculture.[57] On paper, the administrative reorganization of forest resources by the state was a success story of government centralization. By 1910, approximately 88 percent of the 8,803,765 hectares of forested land belonged to the state.[58] However, centralization did not result in improved stewardship of the empire's forests.

The Ottoman forests closest to the sea (such as those on the Gallipoli peninsula) were most seriously compromised in the decades just prior to World War One due to the ease of exporting the wood. The region which included the Aegean islands in particular saw a 60 percent drop in forest coverage between 1897 and 1906.[59] This deforestation problem was not rooted in the lack of legislation or policies, but rather in the inability to enforce the laws that had been created during the many attempts of the Ottomans to practice a more scientifically grounded forestry management. In practice, trees continued to be harvested illegally by local farmers and speculators, and the concessions granted for mining and railways further helped to decimate forests surrounding these concession areas, a result of the need to use wood for smelting and charcoal production. New railways, and particularly the wooden ties for the railroad tracks, which needed to be replaced every five to ten years, had contributed in a major way to deforestation in Thrace and Gallipoli, as in many other parts of the world.[60]

The landscape of Gallipoli just prior to World War One

Today, much of the landscape of the Gallipoli peninsula is covered with thick forests and it is difficult to image how barren the hills once were, as many are now covered with pine trees and other flora, the result of recent reforestation projects after the 1994 fire. Deforestation certainly accelerated during World War One, but even prior to the battles on the Gallipoli peninsula eye-witnesses noted the lack of vegetation. A few days before the fighting began, Liman von Sanders, the German commander of the Ottoman troops, described the landscape in the region: "A few lone pines, a few bushes on the ridges and on the banks of the creeks and coulees, formed the sole natural covering of the generally waste landscape. The cultivation is dependent on the water supply and limited to the surroundings of the few villages, which are all on low ground. It is only the valleys around the small town of Maidos, the latter lying directly on the straits, that show some higher cultivation. Here there are olive and mulberry plantations. Towards the upper part of the Gulf of Saros (Xeros) the interior of the peninsula contains considerable plains which are open and more fertile."[61]

In the final pages of his account of five years in the service of the Ottoman sultan, Liman von Sanders also recommended that the Ottomans would be wise to prepare themselves for future attacks in the region by cutting a canal at the northern entrance to the Dardanelles which would connect the Marmara Sea and the Gulf of Saros, just west of Bolayır, to the Dardanelles. This second waterway would "afford a secure exit to the Turkish fleet, which will always have to be based in Constantinople."[62] It is unclear whether von Sanders had known about the earlier Byzantine plans to radically alter the topography of the Gallipoli peninsula—in this case transforming it from a peninsula, or "*yarımada*" (half island) as it is referred to in Turkish today, to a true island separated from the Thracian mainland. He was aware of the difficulties this suggestion could pose for the Ottomans, who were suffering financially from the costs of the Balkan and then the Çanakkale Wars, but concluded that "its construction offered no technical difficulties as surveyed and attested by experts. Perhaps the project will be resumed at some future time."[63]

People and the pre-war landscape

The region of the Dardanelles was quite an ethnically diverse part of the Ottoman Empire long before the upheavals of the late nineteenth and early twentieth centuries.[64] The lands along the straits and the Gallipoli peninsula were of increasing importance for the Ottoman state from the fifteenth to the seventeenth century as they constituted a gateway to the capital of the empire. The degree to which the Gallipoli peninsula in the early modern Ottoman empire was regarded as a frontier land by the government can be detected in the accounts of Naima, an Ottoman chronicler from the early eighteenth century, who reported that the inhabitants of the village of Seddülbahir were not initially very enthusiastic about the Ottoman Valide Turhan Sultan's project to build a fortification on the peninsula, as there was a worry that the fortress itself would serve as a refuge for bandits who could tyrannize the locals and force them to abandon their homes.[65]

Fortresses and other imperial works in this region of the empire were intended to keep foreign invaders at bay but also to help to make this remote area of the Aegean frontier more habitable. Therefore, along with the "strong encircling walls" of Seddülbahir and the "bastions rising to the sky" that are described in Turhan Sultan's foundation's charter (*vakfiye*) for the outmost fortresses on the Dardanelles, the queen mother also endowed within the fortress of Seddülbahir "a grand mosque to be built for the fortress's guards and wardens . . . along with a fine school, a clean public bath and numerous houses, shelters, shops and markets unlike those found anywhere else."[66] In spite of the efforts of Turhan

Sultan to civilize this frontier land, travelers such as William Grelot were not impressed by the late seventeenth-century settlements on the peninsula. Grelot had documented the fortress of Seddülbahir in his chronicle and made an engraving of it which was published in 1683; he was more impressed by the fortress than by the village of Seddülbahir, which he remarked was "a small village, unremarkable for nothing"[67](see Figure 2.7).

Cook's study of the Troad region and the Dardanelles indicates that by the end of the eighteenth century there was a population of approximately ten thousand living on the lands bordering the straits, which included the Gallipoli peninsula.[68] Data from the Ottoman census of 1831 reveal that the total population of the Biga District, of which the Gallipoli peninsula was a part, consisted of 28,305 Muslims and 7,970 non-Muslims. Archival records from 1840 for the Dardanelles show that a diverse amount of agriculture and animal husbandry was practiced on the peninsula, mostly on small-scale plots of land.[69] As the aforementioned cartographers had also noted, small-scale farming was being practiced at this time on the peninsula. Vineyards, gardens, and olive groves were cultivated, as were oak and chestnut trees. Just prior to World War One, the residents on the peninsula were a multi-ethnic and multi-faith population of Jews, Greeks, Armenians, and Muslim Turks. In 1914, the population of the district of Gallipoli was 28,123 with 17,712 living in Eceabat. Of those living in and around the town of Gallipoli, there were 8,220 registered as Muslim, 16,137 as Greek (*Rum*), 1,190 as Armenian, and 2,576 as Jewish. In the largest town of the peninsula itself, Eceabat/Maydos, there were 7,391 Muslims, 10,292 Greek (*Rum*), 5 Armenians, and 3 Jews.[70]

A landscape transformed by war

Far more invasive contact with the landscape of the Gallipoli peninsula (beyond mapping, deforestation, and farming) occurred once the region was transformed into a battlefield in 1915. This dramatic metamorphosis, and the manipulation of the landscape that accompanied it over the course of two years was well documented on both sides of the trenches by those who fought there and lived in, above, and under the earth, in the tunnels, dugouts, trenches, and fields of Gallipoli. Recent archaeological surveys and excavations on the peninsula are only now revealing the extent to which the forests, landscape, and earth on the peninsula were eradicated or dramatically manipulated to accommodate the many needs of war. In wartime the relationship of humans with the flora, fauna and the earth itself changes as the cultural landscape is transformed into a battlefield landscape, sometimes in quite unexpected ways. Nicholas Saunders, for example, has studied this transformation through the wooden works of art made from trees by soldiers on the Western Front and noted that, "human relationships with Nature, and trees especially, were reconfigured, and wood, leaves, roots, bark, and branches took on new symbolic meanings and sensorial qualities as well as practical purposes."[71]

The transformation in the Gallipoli landscape from 1914 through 1916 has only recently been the topic of academic research. From 2005 to 2014, the ANZAC battlefields of Gallipoli were the focus of the Joint Historical and

Archaeological Survey (JHAS) of the Gallipoli Battlefields project undertaken by the University of Melbourne and the 18 March University in Çanakkale, Turkey.[72] As Antonio Sagona, the main author and editor of *Anzac Battlefield* and leader of the JHAS wrote, "Within eight months Allied and Ottoman forces changed this land, which is actually quite small in size, into a battleground, scarring it with a complex labyrinth of military earthworks."[73] The JHAS project falls into a relatively new sub-discipline of "conflict archaeology," and did not include any archaeological excavation per se but used non-invasive techniques such as field walking, surface inspection, and ground-penetrating radar (GPR) to understand the materiality of the battlefield and what it must have looked like during the war a century earlier, particularly its vast network of trenches and other earthwork types.[74] The project focused on the northern battlegrounds of the ANZAC region only, and among many other features of the wartime landscape the JHAS was able to document in a GIS (geographical information system) database 16.5 kilometers of trenches, 82 tunnels, and over 1,200 artifacts.[75] This important survey has shown that, particularly in the ANZAC region of the peninsula, the degree to which the troops there reshaped the topography of the Gallipoli landscape was astounding.

The Australian soldiers were frequently referred to as "diggers" because of the extensive labor they expended on creating a network of trenches and underground tunnels in that section of the peninsula.[76] C.E. Bean, Australia's official government historian, commented that during the months of conflict that ensued after the 25 April 1915 landing, the amount of actual digging into the earth of Gallipoli, "never ceased, night or day, until the Evacuation. . . ."[77] Digging became a competitive obsession; on 25 July 1915, the men in one battalion "dug 39 feet in just 12 hours, no doubt pressed on not by fear, but by a competitive edge to assert their superior workmanship."[78] The continual movement of earth in the ANZAC sector of the Gallipoli landscape through explosions or tunneling was also noted by Ottoman Major Zeki Bey who recalled in his memoirs written after the war that "I used to watch . . . heaps and heaps of earth always accumulating and extending, everywhere heaps and more heaps . . . I used to say to myself, 'What are they about? These Australians will tunnel to Constantinople'."[79]

The Ottoman troops also dug trenches that made deep and wide cuts into the land, as noted by Ottoman Commander Şefik Aker, who instructed his regiment to construct communication trenches three meters wide and three meters deep. Tunnels were bored into the earth and the Ottomans reinforced them with brick walls and buttresses; they occasionally intersected with those of the Australians, as documented on 20 July 1915 when an Australian tunnel was exploded by the Ottomans, leaving many dead and a crater ten meters wide.[80] As would be expected, while the actual fighting raged on the peninsula, the continued depletion and frequent destruction of natural and agricultural resources on the peninsula that had occurred during the late Ottoman era accelerated. Defensive works of brick and stone were erected, and local timber was harvested from the peninsula and the Kaz (Goose) Mountains on the Asian side. What trees had survived the earlier Ottoman mismanagement of the forests were rapidly consumed for constructing defensive architecture and to provide covers for trenches.[81] Trench and dugout construction, and offensive mining through the peninsula by both Ottomans and Allied forces, manipulated large sections of earth, particularly at the front lines of conflict. As Chasseaud and Doyle have remarked, "On the Anzac front, mining and tunneling activities led to the development of an extensive underground tunnel system which sent forward a series of saps under the enemy lines which were to be packed with explosives and blown at appropriate moments."[82]

The landscape embodied

Embedded in the concept of the cultural landscape and the definitions of historical battlefield terrain is the recognition of strong connections, interdependence, and interaction between humans and their surrounding geographies. Soldiers who fought on the Gallipoli peninsula frequently established personal connections to the actual landscape, as evident in remaining letters, dairies, sketches, and paintings. Sandy Callister has noted that the novice soldier's first experience of Gallipoli was often that of a "touristic gaze," but that the remote experience of the peninsula changed quickly once fighting began and more intimate contact with the landscape occurred."[83] Troops dug into the landscape and their lives became closely connected to the earth, which at times would save their lives, and at other times bury them alive. The British poet John Masefield wrote several evocative descriptions of the Gallipoli battlefield, exploring the complex and intimate relationship of the Allied soldier to the physical landscape of the peninsula once the battles there had commenced. In his literary passages, the topography, dirt, stones, and dust of the peninsula are the embodiment of the soldiers and their struggle to survive. In several instances Masefield writes about specific locations on the battlefields and describes how they were named, some serving as poignant metaphors for the human body. "We had already held a little of the Lone Pine Plateau; our trenches bulged out into it in a convexity or salient known as The Pimple, but the Turks held the greater part, and their trenches curved out the other way, in a mouth, concavity or trap opening towards The Pimple as though ready to swallow it."[84]

Ottomans and Allied forces during the war both likened topographical features of the peninsula to human or mythical bodies, and various names, such as the Boğaz (throat), the Nek, the Pimple, and the Sphinx were used to demarcate significant places on the peninsula. Even memorable days on the battlefield were commemorated with nomenclature evoking the body, such as the snowstorm of 28 November 1915 which was recorded as "Frozen Foot Day" due to the many instances of frostbite that occurred.[85] Specific locations on the battlefield landscape were also given names of familiar locations from home, or of comrades who had died. Descriptive and personal titles given to both the Ottoman and Allied positions attest to the importance of familiarizing and possessing the landscape through the process of naming: Monash Valley, Malone's Terraces, Dead Man's Field, Zimmerman's Farm—or in Turkish, Bloody Ridge (Kanlı Sırt), Valley of Fear (Korku Deresi), Kemal's Place (Kemalyeri).[86]

On some occasions, such as during the disembarkation of the Allied troops on 9 January, the features of the land itself and the temporary graves of the dead served as metaphors for comrades who had been killed and were left behind on the shores of Cape Helles. "By 4 o'clock on the morning of the 9th of January, the last man had passed the graves of those who had won the beaches. They climbed on board their boats and pushed off. They had said goodbye to the English dead, whose blood had given them those acres, now being given back. Some felt, as they passed those graves, that the stones were living men, who cast a long look after them when they had passed, and sighed, and turned landward as they had turned of old."[87]

John Hardgrave, who served with the 32nd Field Ambulance, X Division of the Mediterranean Expeditionary Force, contemplated another scenario for the fallen as he looked out on the Aegean waters below his post: "We deal with the wreckage thrown up by the tide of battle, and wreckage is always a sad sight—human wreckage most of all. But the bay was always full of interest for me, with its ever-changing colour, and the imprint of the ripples in the gleaming silver-sand. And the silver moonlight silvers the silver-sand, while the skeletons of the Xth sink deeper and deeper, to be rediscovered perhaps at some future geological period, and recognised as a type of primitive man."[88]

At the southern end of the peninsula in the area around Cape Helles, the French dug in at the Ottoman fortress of Seddülbahir from 26 April 1915 and stayed until late December 1915. Here too, the connection to the wartime landscape was recorded by those who had survived the landings of 25–26 April 1915. Allied and Ottoman soldiers' diaries kept during the war provide grueling accounts of battles, casualties, burials, and observations of the surrounding landscape and its destruction in this section of the peninsula, but scattered throughout the pages of these memoirs are observations of the flora and fauna of the peninsula. Some soldiers, such as Ottoman Lieutenant Mehmed Fasih, who kept a dairy throughout the fall and winter months of 1915 while fighting at Lone Pine (Kanlı Sırt), provided only brief descriptions of his surroundings. Fasih's entry of 16 November 1915 between 5:00 and 6:00 a.m. gives a glimpse of a few calm moments spent watching the sunrise. "Have some tea. Dawn is breaking. Wash my face and settle down to smoke my water pipe. Daylight gradually spreads and will eventually bathe everything. The valleys become pictures. The hills stick out. Puddles caused by last night's rain shine like mirrors. . . . Purple clouds behind the *Adana* slope are beginning to turn red and pink on their lower edges . . ."[89]

Soldiers also found solace in befriending local wildlife, or tending a vegetable garden (Figure 2.9). The French doctor Joseph Marguerite Jean Vassal, who was part of the medical staff stationed at Seddülbahir from 30 April 1915 to 4 October 1915, wrote poignantly of the unsuspecting fate of the wildlife on the peninsula which were caught in the violence of war, and noted the resilience of soldiers and nature in the face of destruction (Figure 2.10). In the letters he wrote to his wife from the front, Vassal first took note of the storks which had remained in the conflict zone after the Turks had retreated due to the fighting near the village of Kumkale on the shore opposite Seddülbahir. "Over this necropolis two birds kept watch—two storks, probably native to the place, that no noise of battles disturbed. They remained perched on a roof during the quarrel between France and Turkey. But before the end of the day it seems that one of them was killed . . ."[90] An entry on another day recounts the destruction of trees, which must have been quite rare by that stage in the war. "Many shells fell on the cypress wood, which shelters a Turkish cemetery. It was a refuge deliciously cool, restful, and full of poetry. Lately these cypress trees seem to have been the special objective of the Turkish guns. Now their destruction is imminent.

Figure 2.9 Captain Alfred John Shout died on 11 August 1915 after being wounded at Lone Pine. While this photograph of Shout playing with a rabbit was taken outside his tent at Mena camp, near Cairo, it captures the feeling of the quieter moments of the war when soldiers could appreciate the flora and fauna of the Gallipoli battlefields.

Credit: AWM CO2016–1

Figure 2.10 A soldier in his dug-out at Seddülbahir, 1915.
Credit: BnF, Rol. 45659

Like arrows' points towards the sky, these trees with their dark verdure were bathed in the smoke of death."[91]

Once ensconced on the European side of the Dardanelles in his dugout near the fortress of Seddülbahir, Vassal was occasionally able to take great delight in the natural beauties of the peninsula and the changing seasons, even during shelling. "Nature here is so beautiful; the light so pure: It is in a field of marguerites and poppies that the shells are falling, and in which the 'marmites' are digging immense holes . . . Naturally there are no inhabitants here, and not a house standing. Nature is marvelous. It is as beautiful as Provence, and it is Spring!!!" Even the rudimentary earthworks were appreciated for their sensory qualities. "There are real beauties in a dug-out. The platform is made of big branches of pine, to which leaves and cones are still attached. The fresh atmosphere is perfumed with resin. Tufts of poppies are so red that their corollas of blood are full of love and war."[92]

Restoring the landscape

For those who experienced the landscape of the Gallipoli peninsula in the decade immediately after the war ended, impressions varied from the lyrical and poetic to the horrific. Thomas Pemberton, who worked for the Commonwealth War

Graves Commission (CWGC) to establish the cemeteries on the peninsula in the decade after the war, initially romanticized the landscape on the peninsula, claiming that "Man has changed the face of this land but little throughout the ages. The mighty struggle of the past has scarcely left a perceptible scar."[93] A gardener for the CWGC, he became a keen observer of his surroundings and provides an interesting description of the flora and fauna that existed on the peninsula just after the war, including some curious observations such as: "Snakes are to be found, the species growing as much as 8 feet in length. For these the shepherds have to keep a sharp look-out, for they attack the she-goats and suck the milk from their udders."[94]

Both Pemberton and R. Kett, the latter also employed by the CWGC to landscape their cemeteries, started nurseries at Kilye, or Kelia (three kilometers north of Eceabat/Maydos) and elsewhere in the ANZAC and Cape Helles areas (Figure 2.11). The CWGC made decisions at this time about the landscaping of both the cemeteries and the reforestation of the peninsula which involved importing several Antipodean species of plants. Later on page 96 of *Gallipoli To-day* Pemberton exclaims that, "Thousands of young trees are growing in the excellent nurseries at Anzac, Kelia Bay, and Helles, and will be planted out in due course, but not in the cemeteries themselves."

The reforestation of large swaths of the peninsula was done with plants and trees which were not native to Gallipoli but were indigenous to Australia and England, as these would be familiar to the families from the Commonwealth countries who would soon come to visit the graves of their loved ones. Cyril Hughes, one of the members of the British military assigned to administer the affairs of the CWGC on the Gallipoli peninsula after the war, reported enthusiastically to C.E. Bean about the massive number of trees and bushes that were part of the post-war planting campaign on the peninsula. "We are pushing on with the horticultural side of the work and I have a nursery at Kelia and another small one at Anzac and Helles. They are supervised by an office from Kew (London). We are planting out 8,000 trees and shrubs this season and will raise in the nursery about 30,000 of various kinds, including a number from Australia, New Zealand. We find it better to raise them from seed and acclimatize them. . . ."[95]

Pemberton also took delight in anticipating the mark the CWGC would make on the Gallipoli landscape with what might be classified today as invasive species of flora. At the base of one of the Australian commemorative monuments he writes that there is:

> a little shrub which none but a New Zealander would give particular notice. It is a *manuka* tree three feet high. It was tenderly reared from seed in England, and just as tenderly brought out to Gallipoli. After further acclimatization it was planted on Chunuk Bair, and there it has prospered. It was the single instance of the manuka taking firm root and withstanding the Gallipoli winter. Next year, it is hoped, the little white flower will bloom beside those who loved it so well and who rest here.[96]

Figure 2.11 The Commonwealth War Graves Commission's nurseries were in Kelia and established to reforest the peninsula after the devastation of World War One.

Credit: CWGC archives

Peter Slade, an Australian scholar, has remarked on the aggressive planting agenda undertaken by the Australians and CWGC on the peninsula after the war, which included the seeding of their native manuka, *leptospermum scoparium*, at the gravesites of their dead. "Presumably they (the manuka trees) are there to keep the soldiers company. It is difficult to categorize this act . . . Certainly the Gallipoli Peninsula looks very much like parts of the New Zealand coastline, particularly around Wellington, as if it represents an attempt to transpose a social view of a landscape to another geographical site."[97]

Pemberton and Kett's efforts to symbolically mark the land and remember the lives of Australians and New Zealanders who had died were echoed in the actions of surviving Commonwealth soldiers who carried back to their native lands seeds from the flowers and trees of Gallipoli as remembrances of the peninsula where they had fought and lost so many comrades. Gallipoli's pine cones were among the more popular souvenirs that traveled across the world in packages sent home from the front, or in the baggage of departing soldiers. Even today, when there are restrictions on importing invasive species of flora, saplings and seeds from plants growing on the Gallipoli peninsula continue to play a symbolic role in many of the commemorative ceremonies held for the Gallipoli dead across the world.

The desire to return from or bring to Gallipoli a souvenir that will regenerate has extended to samples of the earth itself. Recent examples of commemorative acts by Turks and Australians that use the actual earth of Gallipoli for ceremonial purposes attest to a growing awareness of the value of both the tangible and intangible heritage and "spirit of place" associated with the actual soil of the landscape of the peninsula. During the 2015 visit to Gallipoli of a delegation of Indigenous Australian soldiers to commemorate their Maori ancestors who had fought with the ANZACs, Major Joseph West, a Murrawarri man from Bourke in New South Wales, explained, "Once we get to Gallipoli, we're actually going to do a number of soil collections in and around the battlefield, because we obviously don't know where these soldiers are lying and resting right now, and so we'll be conducting a calling ceremony as part of that . . . In this instance, what we'll be doing is inviting the spirits to wake up and join us on our journey home."[98]

The use of the earth of Gallipoli for ceremonial purposes has also been a part of past and recent Turkish commemorative acts. During the tenth anniversary celebrations in Ankara of the founding of the Turkish Republic soil from regions throughout the country were wrapped in a red flag and presented to Atatürk; in 2014, a 650 kilometer walk from Çanakkale to Ankara was undertaken by İbrahim Batdal, the head of Çanakkale's chapter of the Turkish Association for the Blind, to deliver to Turkey's President Recep Erdoğan a Turkish flag and sample of earth taken from the memorial park of Gallipoli's Turkish 57th Infantry Regiment.[99] In 2016, İsmail Kaşdemir, director of ÇATAB, while praying at a newly discovered grave of an Ottoman soldier on the peninsula, further emphasized the importance of the role that the earth itself plays in the commemorative acts related to Gallipoli. According to Kaşdemir, the martyrs of Gallipoli are the ingredients needed for the sanctity and fortitude of the landscape: "The spirit of Çanakkale is the yeast for this earth. Together, God-willing we will make this earth more powerful."[100]

Reclaiming the landscape

The Armistice was signed on 30 October 1918 and Ottoman and Allied hostilities officially came to an end.[101] Residents of the Gallipoli peninsula began returning to their villages and lands to repair homes and to try to farm the war-ravaged landscape. The terms of the Armistice were intended to maintain order in the region until the remains of the Ottoman Empire could be divided among the Allied powers. It dictated that Ottoman troops were not to be stationed anywhere on the peninsula, so early in 1919 the 12th Senegalese Regiment of the French army returned to the Gallipoli peninsula and were stationed at several locations including, Alçıtepe, and the Ottoman fortresses of Kilitbahir and Seddülbahir.[102] Mithat Atabay notes that during these early days of occupation approximately 5,000 Greek and Armenian recruits were brought to the peninsula by the British to repair the entrenchments and to assist in extending the Decauville railway,

Figure 2.12 Turkish families returned to farm on the peninsula after the war, 1919.
Credit: AWM P07906.025

a narrow-gauge railway that had been constructed by the Allies during the war to transport ordnance and other supplies.[103] The end of the fighting also provided the French, who had been stationed at the fortress of Seddülbahir during the war, an opportunity to resume the archaeological excavations they had initiated at Elaious in the Cape Helles region of the peninsula, and to more thoroughly excavate the so-called "Protesilaus" tomb at the entrance to the village of Seddülbahir. Returning again to the Seddülbahir fortress, the officers of the École Française d'Athènes and their assistants could now concentrate, without the sound of the "marmites," on their archaeological excavations, and on exhuming the bodies of the French soldiers who had been laid to rest in the fortress during the war so that these could be reburied in the main French cemetery at Morto Bay near Cape Helles.[104]

Some villagers who had lived on the Gallipoli peninsula prior to the outbreak of war returned to the peninsula, moving into houses constructed for Greek refugees by the Venizelos government; most Greek Orthodox families had left the peninsula for Greece before or during the 1922 Turkish War of Independence, or did so a few months later during the Population Exchange of 1923.[105] The Convention and Protocol Concerning the Exchange of Greek and Turkish Populations were signed by both governments on 30 January 1923. The third article of the convention addressed provisions and terms of compensation for persons who had left the territories where populations were to be removed after 18 October 1912. According to the terms of the Population Exchange agreement, Orthodox Greeks who had been subjects of the Ottoman sultan were allowed to remain in Constantinople, western Thrace, and the islands of Imbros (Gökçeada) and Tenedos (Bozcaada) just off the shores of the Gallipoli peninsula, but the Greek subjects of the Ottoman Empire who were living on the peninsula itself were to be included in the exchange.

Many Greeks and Turks who had fled or been evacuated from the Gallipoli peninsula just before the outbreak of war never again returned to their homes.[106] The first group of 372 Turks to be brought officially to the peninsula during the exchange arrived on 5 January 1924 from Thessaloniki.[107] More followed from other regions of Greece; 3,788 emigrants were settled on the peninsula between 1924 and 1928 and allocated the homes, fields, shops, vineyards, and animals of those who had left (Figure 2.12).[108]

Statistics of properties lost, gained, and exchanged during these tumultuous post-war years do not reveal the personal struggles that individuals living, or who had once lived, on the peninsula experienced.[109] Several court cases and appeals documented in the CWGC Archives and the National Archives in London attest to the personal and individual anguish of Turks and Greeks caught up in the bureaucracy and politics of those years. Many who fled their homes never did receive compensation for the properties they had vacated prior to the outbreak of war on the peninsula, and repeated legal appeals to Ottoman and British officials often led to no results. This was the case of Mr. Nicholas Papatheodorou who had owned land in Kelia (three kilometers north of Eceabat and also known as Kilye, Koila, and Koela), which he had vacated at the onset of war on the peninsula and then fled to Portiano in Lemnos, Greece. Responding to his appeal of 1 January 1928, in which Papatheodorou claimed that he had not been compensated for his property on the peninsula, the Lieutenant Colonel of the Imperial War Graves Commission (IWGC), the organization responsible at that time for the cemeteries of fallen soldiers of the British Empire, wrote:

> Dear Sir,
>
> Reference your letter of 1 January 1928 regarding property at Kelia previously occupied by you. As you are aware there is an agreement between the Greek and Turkish governments in regard to abandoned property of each nation. The point the Commission must decide is if the land at Kelia can still be considered yours after you left it, or if you have not already been compensated by the Greek government for the land at Kelia.[110]

The Reverend Stylianos Papadopoulos, a Greek priest from Larissa in Thessaly, had become a Greek citizen in 1914 before the war and had owned 1.5 hectares of land on the peninsula. After the war had ended and commemorative building began, he discovered that his land had been allocated to the British to use as a cemetery. In a letter from the British law firm of Leader, Plunket and Leader of 49–50 Newgate Street in London, dated 5 March 1930, the Reverend's case was presented to the CWGC, the organization that had taken over the responsibility from the IWGC for the maintenance of the British cemeteries on the peninsula:

> Although a Greek subject he still owned the property in Kuthia, which was taken for a cemetery for British soldiers. He is unable to get any local evidence, but the hearsay of relatives, and that the British representative appeared to be a Colonel Hughes (an Australian). The land was apparently not given up by the Turkish authorities under Treaty, but apparently just taken and used. Our client's difficulty is that as Greeks were expelled from Chanak and Dardanelles, evidence cannot be obtained by him but doubtless His Majesty's representatives have sufficient power to investigate the matter. If the land was placed at the disposal of the British Authorities by the Turkish Authorities, then as our client was never formally divested of it, they had no right to grant it. Our client as a Greek Priest desires to say that he is honored that the land should be the last resting place of British Soldiers, but respectfully submits that his present situation justifies his asking that some consideration should be given him for the loss of his property.[111]

Claims to repossess property continued for at least a decade after the Population Exchange, and not all cases involving contested property on the peninsula were from disenfranchised Greeks. The case of Kamil Efendi Kadızade, a grocer from the village of Seddülbahir, against Captain Vickery of the British War Cemeteries continued in court for several years, with the plaintiff eventually losing his claim to 6000 square meters of his land and some huts that were on the property. Filed on 19 November 1934, Kamil Efendi Kadızade's appeal states,

> I beg to claim that a part of my land which I possessed near the village of Sedibahir [Seddülbahir], has been, and still is occupied by the British war Cemeteries, since 1921, including four huts on the said land, where a nursery of plants has unlawfully been established. The size of the above-mentioned land is approximately 6 *dönüms* [1 dönüm = approximately 1000 square meters]. Will you therefore please summon Captain Vickery, the Representative of the said War Cemeteries, and after having heard my case make a legal decision that the land be returned to me, also that the rent for 13 years be paid for said land and huts. The amount of the rent could be fixed by a specially formed Committee of Arbitration, and the expenses of the Court and fees of my solicitor must be paid by the abovementioned officer at the Cemeteries. 19–11–1934 Signed: Kamil, The Grocer, Yalı Caddesi No. 61 Çanakkale.[112]

The case made its way through various diplomatic channels to Ankara, and the verdict was eventually decided against Kamil the grocer. The British were assured by the Turkish Ministry for Foreign Affairs in February 1936 that they would not have to cede the land where they had constructed a home for the caretaker of the Cape Helles cemeteries. Nor would they have to compensate Kamil for lost rent as the property should have already been expropriated by the Turkish Ministry of the Interior. Had that not been done properly, it would be.

Descriptions by Pemberton of some of the largest villages of the peninsula—Kilitbahir and Eceabat—are hardly flattering of the landscape or its inhabitants and reveal the struggles and devastation that were faced by those returning home to the peninsula, and its new residents:

> Seen from the Narrows these towns are picturesque. Seen at close quarters there is nothing to commend them . . . Houses have fallen into decay. There has been no effort to keep even the occupied houses in repair. There was a Greek Church of some pretentions, but when the Turks took control again they destroyed it so that not one stone was left upon another. Glass in the windows of the houses is rare . . . Maidos is worse . . . The Greek Government built a number of new houses for their refugees at the south of the vıllage, and it is these which are mainly occupied by the Turks to-day. There is no comfort, no beauty, no picturesque in the houses of peasant Turks.[113]

Conclusion

The still-life paintings done by George Lambert in 1919 of Gallipoli winter wildflowers growing in the post-war ANZAC landscape are evocative of the diverse colors of the flora that Vassal and others had noted during the war (Figure 2.13). Lambert had served in Egypt and Palestine in 1918 and traveled in 1919 to Gallipoli as the official artist for the Historical Mission of the Australian government. His better-known paintings such as *ANZAC, The Landing, 1915,* produced after returning from his trip to the Gallipoli peninsula for the Australian

Figure 2.13 George Lambert's (1873–1930) *Gallipoli Wild Flowers* was painted on 27 February 1919 while Lambert was working with C.E. Bean. On 1 Mar 1919 he wrote: "Again the rain and therefore the account of my day is easy. The flower-piece is finished . . . The flowers are in a biscuit tin on top of a bed for a tent pole; the work is up to standard. Yet, flowers and walking sticks! And I should be painting against time up at the Nek!" Oil on canvas 36 × 46.1 in.

Credit: AWM ART02838

Memorial's exhibition in Melbourne in 1922, were much larger and showed the epic struggle of the Australian troops to take the hills above Arıburnu on the morning of 25 April 1915 (Figure 2.14). Lambert likened the determination of the troops, dwarfed by the immense challenge of scaling the hills, to a swarm of ants.[114] As the former curator of the Australian War Memorial Janda Gooding has remarked, in the painting of the landing Lambert "deliberately painted the bodies and the earth in the same tone, making them nearly indistinguishable. From his own experience walking the ground of Gallipoli, he was only too aware that the earth had literally swallowed up many of these men and obliterated their identities."[115]

This description of the earth devouring its dead was a powerful metaphor also used by the Ottomans to describe the landscape of the war at Gallipoli and the embodied connection to the landscape experienced by the soldiers who had died there. In 1924, Mehmet Akif Ersoy, upon hearing news of the Ottoman victory

Figure 2.14 George Lambert ANZAC, *The Landing 1915* shows Australian troops trying to capture Plugge's Plateau, The Sphinx, Walker's Ridge, and Baby 700. Oil on canvas 190.5 × 350.5 cm.

Credit: AWM ART02873–1

at Gallipoli, wrote "To the Martyrs of Çanakkale", a highly evocative poem, nine lines of which read:

> Earthquakes regurgitate the bodies of the dead!
> Bombs' thunderbolts explode brains,
> Rupture the breasts of brave soldiers.
> Underground tunnels writhe like hell
> Full of the bodies of burn victims.
> The sky rains down death, the earth swallows the living.
> A terrible blizzard heaves men violently into the air.
> Heads, eyes, torsos, legs, arms, chins, fingers, hands, feet . . .
> Body parts rain down everywhere . . .[116]

Two decades later a less violent, but equally iconic and embodied reference was made by Atatürk in 1934. It evoked the landscape of Gallipoli and the actual earth of the peninsula in a poignant description of the relationship of the land to the soldiers now buried within it. Gallipoli was no longer Chronos devouring and regurgitating its children; time had transformed its gender and tone to become a receptive and peaceful Mother Earth, welcoming all her dead sons to her bosom, regardless of nationality.[117]

> Those heroes that shed their blood and lost their lives . . . You are now lying in the soil of a friendly country. Therefore, rest in peace. There is no difference between the Johnnies and the Mehmets to us where they lie side by side here

Figure 2.15 The order "Dur Yolcu [Halt Traveler]" from "Bir Yolcuya [For a Traveler]" is inscribed on the hills of the Gallipoli peninsula at the Kilitbahir ferry crossing to Çanakkale.

Credit: Archive of L. Thys-Şenocak

> in this country of ours . . . You, the mothers who sent their sons from faraway countries, wipe away your tears; your sons are now lying in our bosom and are in peace. After having lost their lives on this land they have become our sons as well.[118]

Inscribed into the landscape behind the military base at the Kilitbahir–Çanakkale Dardanelles crossing (or the "Narrows" as it was referred to by the Allied forces) is an immense epigraphic monument composed of red and white stone plaques embedded into the landscape and large enough to be read from the opposite shore (Figure 2.15). The first few lines of the text are taken from a longer poem of four stanzas entitled "Bir Yolcuya" [For a Traveler], written by Necmettin Halil Onan in 1960.

> *Dur yolcu!*
> *Bilmeden gelip bastığın,*
> *Bu toprak, bir devrin battığı yerdir.*
>
> *Traveler halt!*
> *The earth upon which you tread*
> *Once witnessed the end of an era.*[119]

While only the first two lines of the poem appear on the hillside, coupled with the iconographic imagery of a soldier—rifle in one hand and arm outstretched towards an eternal flame—the "Dur Yolcu" inscription, like a *memento mori*, beseeches those who pass through or live on the Gallipoli peninsula to take a moment to stop, reflect, and honor the sacrifices made on this landscape. It invites those who have memorized the poem, or at least some part of it, to complete its remaining verses, the last stanza of which describes the landscape of the Gallipoli peninsula as made from the blood, flesh, and bones of those who have died upon it and are buried within.

> *Think, the consecrated blood and flesh and bone*
> *That make up this mound, is where a whole nation,*
> *After a harsh and pitiless war, alone*
> *Tasted the joy of freedom with elation.*[120]

In the second and third lines of the poem's first stanza, the most intimate relationship between humans and the Gallipoli peninsula is suggested, and both the tangible and intangible components of this cultural and battlefield landscape are represented by the embodied earth of the Gallipoli peninsula. Here the land asks to be listened to, it breathes as if it is a living creature, and it watches us from a great height, while inviting us to rest our heads on the earth and listen to its heartbeat as we remember the dead:

> *Bend down and lend your ear, for this silent mound*
> *Is the place where the heart of a nation sighs.*

Notes

1 Canada Veterans Affairs, "Vimy Declaration for the Conservation of Battlefield Terrain" (2014); www.veterans.gc.ca/eng/remembrance/memorials/overseas/first-world-war/france/vimy/declaration#Preamble.

2 Ibid.

3 Ibid.

4 Peter Chasseaud and Peter Doyle, *Grasping Gallipoli: Terrain, Maps and Failure at the Dardanelles, 1915* (Staplehurst: Spellmount, 2005).

5 For an extensive discussion of this see Michael Greenhalgh, "French Military Reconnaissance in the Ottoman Empire during the Eighteenth and Nineteenth Centuries as a Source for Our Knowledge of Ancient Monuments," *The Journal of Military History* 66 no. 2 (2002), 359–88.

6 Greenhalgh, 368.

7 Chasseaud and Doyle, 21.

8 For general overviews of Ottoman cartography in English and Turkish see Kemal Özdemir, *Ottoman Cartography*, (İstanbul: Avea, 2008); see also Doğan Uçar, "Turkish cartography in the 16th century," *Science and Civilization*, ed. E. İhsanoğlu and F. Günergun (İstanbul: IRCICA, 2000); *Üç Denizin Arasında: Osmanlı ve Fransız Boğaz Haritaları/Entre Trois Mers: Cartographie Ottomane et Française des Dardanelles et du Bosphore (XVII-XIX siècle)* (İzmir: Arkas Holding, 2016).

A recent study of maps of the Dardanelles in Turkish is by Y. Koca, "Dardanelles, Hellespont, Çanakkale: Çanakkale Haritaları," *Çanakkale Araştırmaları Türk Yıllığı*, 22, no. 15, (2017), 109–34.

9 For a useful source in Turkish about the eighteenth through early twentieth-century defense system of the Dardanelles see, Salih Hatip, "Çanakkale Boğazı (Bahr-i Sefid Boğazı) Savunma Sistemi 1770–1918," unpublished Ph.D. diss., (Hacettepe University, 2013); *Üç Denizin*, 163–5.

10 The Bonne projection is a pseudo-conical map projection; Chasseaud and Doyle, 148.

11 For a study of Mehmet Şevki Pasha's maps see *Çanakkale Tahkimat Haritası/ Defense Atlas of Çanakkale 1916*. Ankara: Genelkurmay ATASE Başkanlığı Yayınları, 2009).

12 See also Edip Özkale and Mustafa Rıza Şenler. *Haritacı Mehmet Şevki Paşa ve Türk Haritacılık Tarihi 1919 Yılına Kadar* (Ankara: Harita Genel Müdürlüğü, 1980).

13 Details can be found in the 2016–2020 Strategic Planning Document produced by ÇATAB available at http://catab.kulturturizm.gov.tr/Eklenti/50049,2016–2020-stratejik-planpdf.pdf?0.

14 Chasseaud and Doyle, 50.

15 Ibid., 52–3.

16 This was the Naval Intelligence Division: (N.I.D). 838 Turkey. *Coast Defences and Resources, Coast Defence Ordnance and Arsenals*, May 1908, The National Archives (TNA) ADM (Admirality) 231/49 and TNA CAB (Cabinet Office) 19 and *Report on the Defences of Constantinople* (1909) from the War Office.

17 Walter Leaf, *Troy: A Study in Homeric Geography* (London: Macmillan and Company, 1912).

18 Chasseaud and Doyle, 20.

19 For an account of von Sanders, five years on the Gallipoli peninsula see Liman von Sanders, *Five Years in Turkey*, (Philadelphia: The Williams and Wilkins Company, 1928).

20 *11th Division War Diary*, TNA CAB 45/258; Chasseaud and Doyle, 248.

21 Chasseaud and Doyle, 241.

22 Chasseaud and Doyle, 241, 247.

23 *Military Report on Eastern Turkey in Europe 1905*, TNA FO 881/8589; Chasseaud and Doyle, 273.

24 This lack of interest in Ottoman forests appears to be changing with some new publications in this field. See Selçuk Dursun, "*Forest and the State: History of Forestry and Forest Administration in the Ottoman Empire*," unpublished Ph.D. diss., (Sabancı University, 2007). See also Çevre ve Orman Bakanlığı (Environment and Forest Ministry), *Osmanlı Ormancılığı İlgili Belgeler*, 3 vols. (Ankara: Çevre ve Orman Bakanlığı, 1999–2003); Yücel Çağlar, *Türkiye Ormanları ve Ormancılık* (İstanbul: Iletişim Yayınları, 1992); Halil Kutluk, ed., *Türkiye Ormancılığı ile İlgili Tarihi Vesiklar*, 1487–1923 (İstanbul Osmanbey Matbaası, 1948). For a history of forests in the larger Mediterranean during the Antique through early modern periods see Russell Meiggs, *Trees and Timber in the Ancient Mediterranean World* (Oxford: Clarendon Press, 1982). For the most comprehensive study of Egypt during the Ottoman era see Alan Mikhail, *Nature and Empire in Ottoman Egypt: An Environmental History* (Cambridge: Cambridge University Press, 2011).

25 Benjamin H. Isaac, *The Greek Settlement in Thrace until the Macedonian Conquest*, (Leiden: Brill, 1986), 186.

26 For an early study of trees and plants referred to in the works of Homer see Edward S. Forster, "Trees and Plants in Homer," *The Classical Review* 50, no. *3* (1936), 98–9.
27 Ibid.
28 Isaac, 160.
29 Ibid.
30 Ibid., 192–3.
31 Ibid., 177.
32 Göksel Sazcı, "Maydos Kilisetepe Settlement Mound," in *City of Legends and Epics Çanakkale*, ed. F. Özden (İstanbul: Yapı Kredi Press, 2012), 227–30.
33 Lucienne Thys-Şenocak, "Divided Spaces, Contested Pasts: The Gallipoli Peninsula Historical National Park," in *Public Nature: Scenery, History and Park Design*, ed. Ethan Carr, Shaun Eyring, and Richard Guy Wilson (Charlottesville: University of Virginia Press, 2013), 130.
34 Kritovoulos, *The History of Mehmed the Conqueror*, trans. C.T. Riggs (Princeton: Princeton University Press, 1954), 197–8.
35 Dejanirah Couto, "Les fortifications ottomanes des Dardanelles/Çanakkale Boğazı' ndaki Osmanlı Hisarları/Les Fortifications Ottomanes des Dardanelles," *Üç Denizin Arasında,* 105–77.
36 For a detailed history of these fortresses see Lucienne Thys-Şenocak, *Ottoman Women Builders: The Architectural Patronage of Hadice Turhan Sultan* (Aldershot: Ashgate, 2006), 108.
37 Süleymaniye Library 150 folio 15a
38 Thys-Şenocak, *Ottoman Women Builders*, 112–13.
39 Asuman Baytop, *Türkiye'de Botanik Tarihi Araştırmaları* (Ankara: Türkiye Bilimsel ve Teknolojik Araştırma Kurumu, 2005), 38.
40 Aaron Hill, "A Full and Just Account of the Present State of the Ottoman Empire in All its Branches: With the government and policy, religion, customs, and way of living of the Turk, in general" London, 1709, *Eighteenth Century Collections Online*, 30.
41 Dursun estimates that in the eighteenth century a single Ottoman three-deck galleon of 52.5 meters required 48,244 pieces of timber. Dursun, 49–50.
42 Alan Mikhail, *Nature and Empire in Ottoman Egypt: An Environmental History* (Cambridge: Cambridge University Press, 2011), 148.
43 Dursun, 65.
44 Ibid., 55.
45 For this document see BOA MMD 6/601 (11 Jan 1565); *Osmanlı Ormancılığı ili İlgili Belgeler I*, 12–13.
46 BOA MD 5/691 (21 Jan 1566); Dursun, 73.
47 Dursun, 211-12. For a summary of terms used in Ottoman forestry management and important dates of legislation see S. Özden and Ü. Birben. "Ottoman forestry: socio-economic aspect and its influence today", *Ciência Rural*, 42 (3) (2012), 459–66.
48 Dursun, 173. It was not uncommon for French experts to serve in the Ottoman administration as demonstrated by Christine Isom-Verhaaren, "Shifting Identities: Foreign State Servants in France and the Ottoman Empire," *Journal of Early Modern History* 8 no. 1 (2004), 109–34.
49 Dursun, 177.
50 George P. Marsh, *Man and Nature*, ed. D. Lowenthal (Cambridge, Mass: Belnap Press, 1965), 9. Marsh had spent over three years in the Ottoman Empire, from 23 February 1850 to 30 November 1853, and had been appointed by U.S. President

Zachary Taylor as the American Minister to the Ottoman Empire during the reign of Sultan Abdülmecid. Marsh was a keen observer of nature as well as the author of several books about the importance of conservation. For an account of Marsh's time in the Mediterranean and Turkey see Lowenthal, *George Perkins Marsh: Prophet of Conservation* (Seattle: University of Washington Press, 2015). Expertise on forestry management and nature conservation would be sought by the Turkish government from the United States many decades later in the 1960s when the Turkish Forestry Ministry worked with the United States National Park Service to establish the first national parks in Turkey.

51 Louis Tassy, *Études sur l'aménagement des Forêts* (Paris: Bureau des Annales Forestièrs, 1858); Dursun, 181–3.

52 Dursun, 186–7.

53 Ibid., 193–4.

54 Ibid., 216–17.

55 Hyde Clarke, "The Right of Property in Trees on the Land of Another, as an Ancient Institution," *The Journal of the Anthropological Institute of Great Britain and Ireland* 19 (1890), 311–79.

56 Kerim Yund, *Türkiye Orman Umum Müdürleri Albümü* (İstanbul: Hüsnütabiat Matbaası, 1959), 11–18.

57 Ibid., 93–101.

58 Dursun, 240.

59 Dursun, 292–3; see also Tevfik Güran, *Osmanlı Devleti'nin İlk İstatistik Yıllığı 1897/The First Statistical Yearbook of the Ottoman Empire 1897*, vol. 5 (Ankara, Devlet İstatistik Enstitüsü, 1997), 177.

60 This was certainly the case in the United States as pointed out by Donald Pisani, "Forests and Conservation, 1865–1890," *The Journal of American History*, vol. 72, no. 2 (Sept., 1985), 344.

61 Liman von Sanders, *Five Years in Turkey* (Philadelphia: The Williams and Wilkins Company, 1928), 58–9.

62 Ibid., 104–5.

63 Ibid.

64 For data on the Gallipoli region in the sixteenth century see Mustafa Kulu, "1530'da Çanakkale Yöresinde Yerleşme ve Nüfus," *Çanakkale Araştırmaları Türk Yıllığı* 2 no. 2 (2004), 220–4.

65 L. Thys-Şenocak, *Ottoman Women Builders*, 114.

66 Süleymaniye Library 150 folio 15a.

67 Guillaume-Joseph Grelot, *A Late Voyage to Constantinople* (London: J. Playford, 1683), 15–16.

68 John Manuel Cook, *The Troad: An Archaeological and Topographical Study* (Oxford: Oxford University Press, 1973), 53.

69 According to Cook in 1840, there were 6,860 acres of cultivated fields (*mezra*).

70 For these statistics see Mithat Atabay, *Atatürk ve Çanakkale* (Edirne: Paradigma Akademi Yayınları, 2015), 207.

71 N. Saunders, "Bodies in trees: A matter of being in Great War landscapes," *Bodies in Conflict: Corporeality, Materiality, and Transformation*, ed. N. Saunders and P. Cornish (London: Routledge, 2014), 22.

72 For the most recent publication on this project, see *Anzac Battlefield: A Gallipoli Landscape of War and Memory*, ed. A. Sagona et al. (Cambridge: Cambridge University Press, 2016).

73 Antonio Sagona, "Introduction," *Anzac Battlefield*, 1.

74 For conflict archaeology see Nicholas J. Saunders, *Killing Time: Archaeology and the First World War* (Stroud: Sutton Stroud, 2007); Antonio Sagona and Jesse Birkett-Rees, "Battlefield Archaeology: Gallipoli," *Anzac Battlefield*, 83–97.

75 Ibid.

76 For a study in the role of ANZAC digging in Gallipoli see Nathan Wise, " 'Dig, Dig, Dig, Until You Are Safe:' Constructing the Australian Trenches on Gallipoli," *First World War Studies* 3 no. 1 (2012), 51–64.

77 Ibid., 58.

78 Ibid., 60.

79 Richard Reid and Ian McGibbon, "Forming the Anzac Battlefield," *Anzac Battlefield*, 125.

80 Mithat Atabay et al., "Forming the Ottoman Battlefield," *Anzac Battlefield*, 157.

81 Peter Doyle and Matthew R. Bennett, "Military Geography: The Influence of Terrain in the Outcome of the Gallipoli Campaign, 1915," *The Geographical Journal* 165, no. 1, (1999), 12–36.

82 Chasseaud and Doyle, 257.

83 Matthew Cunningham here quotes S. Callister's *Face of War* in his article, " 'Familiarising the Foreign:' New Zealand Soldiers' Observations On Landscape During The Gallipoli Campaign," *New Zealand Journal of History* 45 no. 2 (2011), 211.

84 John Masefield, *Gallipoli* (New York: Macmillan, 1917), 158.

85 Ibid., 228.

86 Antonio Sagona, et al. "Gallipoli Battlefield Place Names," *Anzac Battlefield*, xxiii–xxvi.

87 Masefield, 240.

88 John Hardgrave, *At Suvla Bay: Being the Notes and Sketches of Scenes, Characters and Adventures of the Dardanelles Campaign* (New York: Houghton Mifflin Company, 1917), 138.

89 Lt. Mehmed Fasih 5th Imperial Ottoman Army Gallipoli 1915, *Gallipoli 1915 Bloody Ridge (Lone Pine) Diary* (İstanbul: Denizler Kitabevi, 1997), 91.

90 Joseph Marguerite Jean Vassal, *Uncensored Letters from the Dardanelles* (London: W. Heinemann, 1916), 112–15.

91 Ibid., 156.

92 Ibid., 83, 91, 108–9 (Entries dated 27 May 1915 and 14 July 1915).

93 Thomas James Pemberton, *Gallipoli To-Day* (London: E. Benn, 1926), 23. Pemberton, who was working on the peninsula for the CWGC, wrote his account of life there eleven years after the major battles were fought.

94 Ibid., 58.

95 AWM38 3DRL 8042/51 (Letter from C. Hughes to C.E. Bean December 29, 1922).

96 Pemberton, 97.

97 "Champion Trees in Commonwealth War Cemeteries"; www.cwgc.org/admin/files/CHAMPIONTREES.pdf. Peter Slade has commented on the exportation of pine trees from the Gallipoli battlefields to Australia and New Zealand, and questions the claim that seeds from the Lone Pine, one of the primary battlesites of the peninsula, were taken back by soldiers to be planted throughout Australia, stating that the last tree in that area was felled on 6 August 1915 by the Turks for use as charcoal. Peter Slade, "Gallipoli Thanatourism: The Meaning of ANZAC," in *Annals of Tourism Research* 30, 4 (2003), 790–1; Peter H. Hoffenberg, "Landscape,

Memory and the Australian War Experience, 1915–18," *Journal of Contemporary History*, 36, no. 1 (January 2001), 125.

98 Lindy Kerin, "Indigenous Soldiers Head to Gallipoli for Calling Ceremony," ABC News, (29 July 2015). www.abc.net.au/news/2015–07–29/indigenous-soldiers-travel-to-gallipoli/6656978. A. Becker notes that the practice of bringing back soil from battlefield pilgrimages of World War One was a widespread phenomenon: "And they brought back a little handful of earth, enclosed in an urn, and placed it at the foot of certain monuments. The earth of Verdun or of the Somme permitted places to be consecrated that had experienced the war mainly as a long litany of local men killed and longed for. Pilgrimages have always been accompanied by the purchase of souvenirs, and shrewd merchants quickly mass produced little terracotta urns in the form of milestones from the "sacred road" to Verdun, complete with helmets and palms, bearing the inscription, 'This milestone contains a bit of sacred earth from Verdun.' " "From Death to Memory: The National Ossuaries in France after the Great War," in *History and Memory*, vol. 5, no. 2 (Indiana University, Fall–Winter 1993), 32–49. For a definition of the heritage concept "spirit of place" see http://whc.unesco.org/uploads/activities/documents/activity-646-2.pdf and www.dva.gov.au/sites/default/files/files/publications/commemorations-war-graves/order-of-service.pdf.

99 Esra Özyürek provides an account in *Nostalgia for the Modern*, of the second day of these anniversary celebrations and the way in which the references of blood mixed with soil were used to tie the different populations of Turkey to the homeland., 131. See also, "Çanakkale'den Getirilen Toprak ve Bayrak Cumhurbaşkanı Erdoğan'a Sunuldu," (15 October 2014). www.tccb.gov.tr/haberler/410/1430/canakkaleden-getirilen-toprak-ve-bayrak-cumhurbaskani-erdogana-sunuldu.html.

100 "Çanakkale ruhu bu toprakların mayasında vardır. Çanakkale Ruhuyla beraber inşallah biz bu topraklarda daha güçlü olacağız." *Anadolu Gazetesi*, 2 February 2017. www.anadolugazetesi.net/mezari-102-yil-sonra-gun-yuzune-cikti-67072.html. Collecting and gifting soil from different regions of Turkey was also done during the 10th anniversary of the Republic Day ceremonies of 1933 in Ankara. See Esra Özyürek, *Nostalgia for the Modern: State Secularism and Everyday Politics in Turkey* (London and Durham, N.C.: Duke University Press, 2006), 126–32.

101 For details about this agreement see Erik Zürcher, "The Ottoman Empire and the Armistice of Mudros," in *At the Eleventh Hour: Reflections, Hopes, and Anxieties at the Closing of the Great War, 1918*, ed. Hugh Cecil and Peter H. Liddle (London: Leo Cooper, 1998), 266–75.

102 M. Atabay, *Atatürk ve Çanakkale*, 107–8 and T.C. Başbakanlığı Cumhuriyet Arşivi BCA 030.00.00.10.54.356.6.1. and 030.00.00.10.54.358.1.1 and 030.00.00.10.54.358.1.2; Pemberton, 38.

103 M. Atabay, *Atatürk ve Çanakkale*, 107–8.

104 See Chapter 3 of this book for a longer discussion of the French excavations in the Cape Helles region during and after World War One.

105 For information on the regional impact of the Population Exchange see J. McCarthy, *Muslims and Minorities: The Population of the Middle East and the Balkans* (Isis Press: İstanbul, 2002), 41–63.

106 Mithat Atabay, "Mübadelede Çanakkale," in *Çanakkale, I: Savaşı ve Tarihi* (İstanbul: Büyükşehir Belediyesi Kültür ve Turizm Daire Başkanlığı, 2006), 605–11. Mithat Atabay, "Cumhuriyet Döneminde Çanakkale'ye Göçler," in *Çanakkale Tarihi*, ed. Mustafa Demir, 6 (İstanbul: Değişim Yayınları, 2008), 3315–32.

107 Atabay, "Mübadelede Çanakkale," 607.
108 Ibid.
109 For more detailed statistics see Atabay, "Mübadelede Çanakkale," 607–8; also Atabay, "Cumhuriyet Döneminde Çanakkale'ye Göçler," 3319–20.
110 The National Archives Foreign Office (TNA FO) 2661.
111 CWGC WG 920/2.
112 CWGC WG 1420.
113 Pemberton, 31–32.
114 George W. Lambert, *Autobiography of George Lambert*, Mitchell Library, Sydney (ML MSS A1811), 75.
115 Janda Gooding, *Gallipoli Revisited: In the Footsteps of Charles Bean and the Australian Historical Mission* (Melbourne and London: Hardie Grant Publishing, 2009), 209.
116 For this translation of Mehmet Akif Ersoy's "To the Martyrs of Çanakkale" by Michael Burch see Şafak Horzum, "Reflections on the Gallipoli Campaign in Turkish Literature," in *The Gallipoli Campaign: The Turkish Perspective*, ed. Metin Gürcan and Robert L. Johnson (London and New York: Routledge, 2016), 215.
117 Translation of this saying according to U. İğdemir, in *Atatürk ve Anzaklar* (Ankara: Türk Tarihi Kurumu, 1978).
118 www.awm.gov.au/encyclopedia/ataturk/
119 *Major and Mrs. Holt's Battlefield Guide to Gallipoli,* (South Yorkshire, Leo Cooper, 2000), 58.
120 The original Turkish version of this poem can be found at www.anzacsite.gov.au/2visiting/turkish_stop.html.

3 Archaeology and artifacts of the Gallipoli peninsula

> I have seen some Greek excavations. In digging trenches soldiers have found stone sarcophagi which we think are more than 2000 years old. The walls are thick and the coffins very big. They are filled with fine earth which has slowly slid through the cracks of the lid. Among this soft earth are bones, quite recognisable, vases, bowls, lamps, sometimes statues. There are also potteries with faces and figures of women . . .[1]

The archaeological and museological present

Given Gallipoli's location on the shores opposite Troy, the past century has seen remarkably few archaeological excavations on the peninsula; more common were discoveries made during military campaigns, such as the one described in Chapter 2 by Vassal, which was undertaken by French soldiers digging for defense purposes during World War One. Evidence of continuous human occupation on the peninsula exists as early as the Neolithic era (mid-7th millennium BCE), and the strategic geographic location of the peninsula between Thracian Europe and Asia, along with its fertile agricultural lands, has ensured its occupation by many different peoples through the ages. Much of the peninsula's history has yet to be discovered since archaeologists were unable to conduct extensive archaeological research in this region for nearly a century after the end of World War One due to restrictions imposed by the Turkish military.

Apart from a few years of illegal archaeological excavations in the late nineteenth century by the famed German archaeologist Heinrich Schliemann, and brief archaeological campaigns undertaken during and shortly after World War One by the French Expeditionary Forces, the Gallipoli peninsula was effectively off-limits to archaeologists and maintained as a Turkish military zone until 1979. Archaeological excavations on the peninsula began again in 2005, first at the Ottoman fortress of Seddülbahir as part of a restoration project conducted jointly by Koç University, Istanbul Technical University, and the Çanakkale Museum of Archaeology.[2] Archaeological research on the peninsula has continued over the past decade and there are currently a limited number of archaeological surveys and excavations at the sites of Eceabat (Kilisetepe), Çamburnu, and the Kilitbahir,

Seddülbahir, and Biga fortresses.[3] Non-invasive archaeological surveys of the peninsula have been conducted for different eras and by various archaeologists, beginning in 1979. This growing interest in archaeological research on the peninsula, the management challenges of these excavations for the local museums, and the implications that new archaeological findings have for our understanding of the landscape of Gallipoli are discussed at greater length in the final chapter, but some aspects of these issues appear here.

Early archaeological campaigns on the Gallipoli peninsula

Numerous references to the ancient past and topography of the peninsula can be found in the works of Herodotus, Thucydides, Strabo, Xenophon, and others. These authors describe events that occurred on the peninsula during its Classical and earlier historical eras, but historians, philologists, and archaeologists have tended until quite recently to focus their archival research and conduct their excavations on the better-known sites of the opposite shore, such as Troy or Alexandria Troas.[4] In 1882 Schliemann began a campaign on the Gallipoli peninsula to find the alleged tomb of Protosilaus, the first of Agamemnon's soldiers to perish in the Trojan War.[5] From his study of ancient writers, Schliemann surmised that the tomb of this mythic warrior had to have been located on Cape Helles, near the entrance to the Ottoman fortress and village of Seddülbahir. He crossed over the Dardanelles to explore what today is the site of Karaağaçtepe, a tumulus now known to have had levels dating back to the Neolithic period (Figure 3.1). Schliemann had little interest in the earlier periods of occupation that predated the heroes of Troy, and was in a hurry to find evidence to support his theories about the legitimacy of Homer's epics and their mythical characters. He did observe, as he burrowed many meters into the mound, which he estimated was 122 meters high and 10 meters wide, that there were similiarities in the ceramics found at this site on the peninsula and Hisarlık where Troy is located. "I was amazed to find not only the tumulus, but also the gardens around it strewn with fragments of thick lustrous black pottery . . ." Schliemann notes that he stored the excavated finds from this mound in one of his wooden barracks at the Hisarlık site and shared others with the Imperial Museum at Constantinople. The people who once lived near Seddülbahir, he concluded, were "of the same race and culture as the first settlers on the hill of Hisarlık."[6] His report on his two-day expedition gives a sense of how the Ottoman government kept a watchful eye on the attempts by Europeans to gather information of any kind, archaeological or otherwise, on the Gallipoli peninsula. Considering the number of archaeologists who had also been involved in gathering military intelligence during the years leading up to World War One,[7] the suspicion of the Ottomans seems warranted. As Schliemann explained,

> . . . I did not lose a moment of my precious time, and having brought with me pickaxes, shovels, baskets, etc. I at once ordered the four workmen to sink, just in the middle of the summit, a shaft three meters in length and

> breath. I had done exceedingly well to hurry on the work, for the commandant of the fortress of Seddul Bahr reported my doings to the military governor of the Dardanelles, who not being able to conceive how a man could waste his time excavating a lonely hillock, suspected that I was merely using the excavation of the Protosilaus tomb as a pretext for making plans of the fortress of Seddul Bahr, and investigating the lines of torpedoes recently sunk in the Hellespont; and so he issued an order to suspend the excavation.[8]

A few decades after Schliemann's escapade, and just prior to the outbreak of war on the peninsula, Charles Picard and A.J. Reinach (the former went on to serve as the director of the École Française d'Athènes from 1914 to 1918, and again from 1919 to 1925) conducted an archaeological survey of the Chersonese de Thrace, and sketched the remains of antiquities in the villages they explored on the peninsula such as Sestos and Maydos (Eceabat) (Figures 3.2 and 3.3).[9] From July through August 1910 they documented in their notebooks the archaeological remains on the Gallipoli peninsula while traveling south from Sestos to Seddülbahir, and suggested that the area of Eski Hisarlık could be the ancient city of Elaious[10] (Figure 3.4). The findings from both the informal and professional archaeological excavations conducted in 1915 on the Gallipoli peninsula by different members of the Allied forces who were stationed there are reminders that, in addition to being a place where European soldiers engaged in combat with the enemy, the landscape of the peninsula—particularly for those who had been brought up reading the works of Homer and Classical Greek mythology—was a place which fascinated the amateur as well as the trained archaeologist-soldiers fighting there.[11] Several of the ancient objects unearthed on the peninsula at this time of upheaval found their way into the pockets of soldiers; some were shipped home to private collections or ended up in established museums such as the Louvre and the Istanbul Archaeology Museum.[12]

Sergeant-Major R.S. Jones's wartime discovery of a Greek dedicatory inscription to Attalus II, the king of Pergamon, by second century BCE inhabitants

Figure 3.1 Tomb of Protosilaus (Karaağaçtepe) as depicted by Schliemann in *Troya: Results of the Latest Researches and Discoveries on the Site of Homer's Troy* (London: John Murray, 1884).

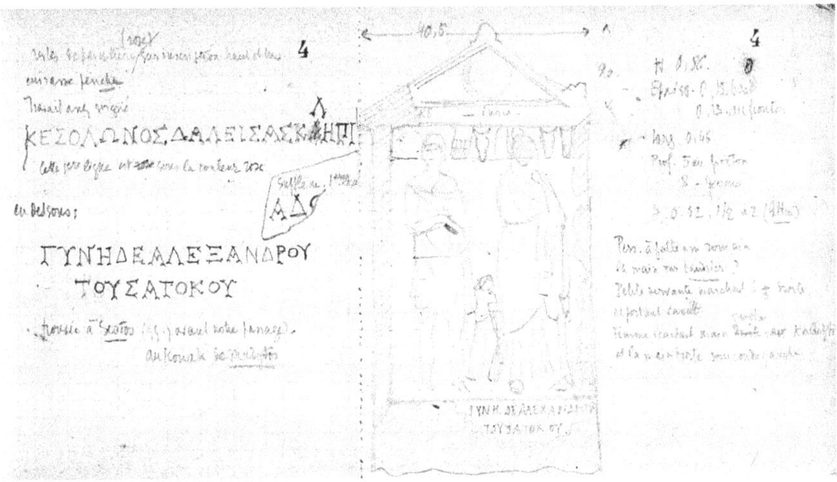

Figure 3.2 Sketches of remains of antiquities in Sestos by Reinach and Picard.

Credit: "Carnet du voyage d'A. Reinach et Ch. Picard en Thrace," 1910. © EFA, inv. THRACE 2"

of ancient Elaious, was made not far from his base at Cape Helles.[13] Captain George Augustus Auden found time between providing medical help to the wounded to investigate archaeological remains, inscriptions, and even a sarcophagus in the Suvla Lake area of the peninsula, an effort which led to the identification of that site as ancient Alopekonnesos and the existence of Koila (or Coela), a Roman site located approximately three kilometers north of today's Eceabat and inland from Kilye Bay where the CWGC had one of their nurseries.[14] The location of the Lone Pine memorial in the ANZAC region today commemorates the soldiers from New Zealand and Australia who went missing in action in this part of the peninsula, but prior to its construction just after World War One, ceramic remains found by sapper Sergeant Lawrence during the war, and Colonel Crouch's post-war reports describing Roman coins and architectural remains, suggest that this area once served as a Roman camp or lookout post.[15]

Archaeology as it was practiced by the amateur or professional was a way to pass time and engage in an activity which uncovered earlier layers of the lands for which they were fighting. From a letter written on 12 June 1915 to his wife by Joseph Marguerite Jean Vassal (the French medical officer stationed at Cape Helles—whom we met in the previous chapter and the introductory quote of this one) it is clear that archaeological remains in that section of the peninsula were abundant and their worth and beauty as antiquities clearly recognized:

> As a diversion from the "marmites" we continue our excavations. We are on a Greek necropolis of the highest antiquity—some five or six centuries before our era. In digging trenches we come on enormous stones which resound. They are the lids of tombs. With great care (but not always) we remove the

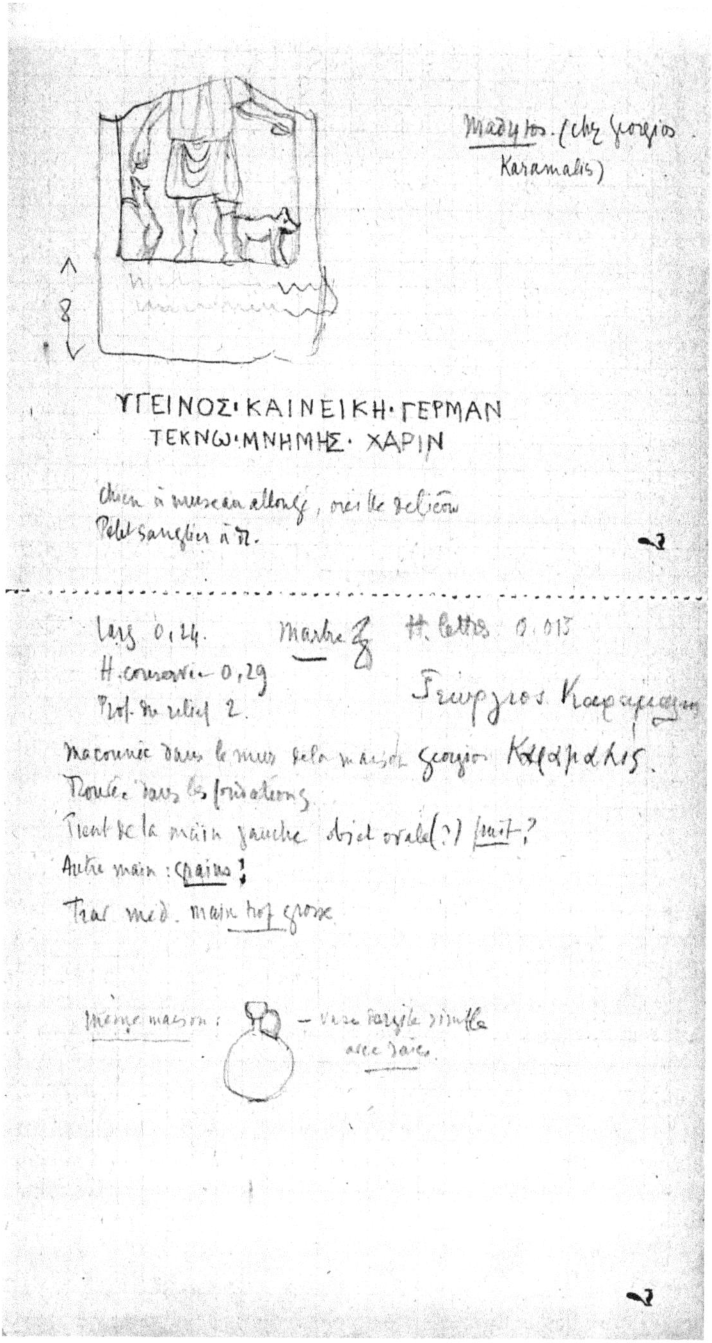

Figure 3.3 Sketches of remains of antiquities in Maydos by Reinach and Picard.

Credit: "Carnet du voyage d'A. Reinach et Ch. Picard en Thrace," 1910. © EFA, inv. THRACE 2"

Ruines de la Ville d'Eléonte.

Figure 3.4 Remains of the ancient city of Elaious, in Gelibolu Peninsula (Ruines de la ville d'Eléonte). From Choiseul-Gouffier, Gabriel *Voyage pittoresque de la Grèce*, Paris, J.-J. Blaise, 1822.

Credit: Aikaterini Laskaridis Foundation Library

> covering stone. Underneath is the interior of a stone coffin which we empty little by little. Grain by grain for centuries the soil has gently penetrated. Inside there is a skeleton more or less preserved. Besides this in each sarcophagus there are potteries more or less rich, but always of exquisite form. I have beneath my eyes a delicate cup which the slightest shock would break. It symbolises that peculiar and characteristic beauty of form which Greece revealed to us. Its long handles, almost ethereal in their delicacy, give to this little thing the palpitations of wings. Only passionate admirers of the human body were capable of creating such marvellous lines in clay. They are arms which open towards the heaven—these handles of my Greek cup. I very much enjoy drawing these treasures. Last night the guns were silent. A respite for all. It is so rare. Good night, Beloved. J[16]

Another letter penned by the same author describes Elaious (or what the French referred to as Eléonte), the site of the archaeological excavation conducted by members of the French military posted at the fortress of Seddülbahir between 8 July and 12 December 1915 and supervised by a succession of archaeologists, most working under the auspices of the École Française d'Athènes: Sergeant Édouard Dhorme, Joseph Chamonard, Fernand Courby, and Lieutentant Leune.[17] The actual work of digging was undertaken by the Senegalese conscripts in the French army[18] (Figure 3.5). It was a well-organized and swift campaign lasting six months and many objects were excavated.[19] The artifacts from these excavations were extensive, ranging from the eighth century BCE to the Roman era.

Figure 3.5 Senegalese conscripts assisted with the excavations of the Protosilaus mound and Elaious during and after World War One.

Credit: Robert Demangel, *Le Tumulus Dit de Protésilas* (Paris: de Broccard, 1926)

At the site of Elaious alone there were 709 tombs, containing 390 pithoi and 319 stone sarcophagi. Funerary jars, vases, amphoras, jewelry made of bronze, silver, glass paste, and shell, terracotta busts and statues were all part of the trove removed from the site[20] (Figures 3.6 and 3.7).

The doctor, Vassal, whose dugout was close to the French excavation site and therefore afforded him the opportunity to watch its progress, revealed in one of his letters home a certain remorse over how the excavation being conducted by his compatriots was disturbing the human remains discovered in some of the tombs at Elaious:

> The excavations of which I have spoken to you, and which contain such beautiful pottery, come from a Greek colony situated on the site of Eléonte [Elaious]. It was there that Miltiades prepared his expedition to Lemnos and that Alexander the Great embarked for the Plain of Troy . . .[21] While digging close by, two enormous earthen jars were found, about one and a half yards long and with an opening of half a yard across. They were side by side, one broken and the other intact. A soldier told me that in each jar there were two skeletons. Man and wife? Lovers? They had slept together for twenty centuries. What profanation, what a crime to have interrupted a union which promised to be immortal![22]

This sentiment to leave the objects in situ does not seemed to have been shared by the archaeologists appointed to the project. Approximately 1,500 objects were removed from the site by the French team; 1,200 of these were sent to the Louvre

Figure 3.6 Images of diverse vases from Elaious now in the Musée du Louvre, Paris Catalog numbers: Elé177; Elé178; Elé179; Elé182; Elé185; Elé186; Elé180; Elé181; Elé184; Elé188; Elé187; Elé190; Elé312.

Credit: Photo © RMN-Grand Palais (Musée du Louvre)/Tony Querrec

Figure 3.7 Tombs at Elaious.

Credit: Bulletin de Correspondance Hellénique XXXIX 1915

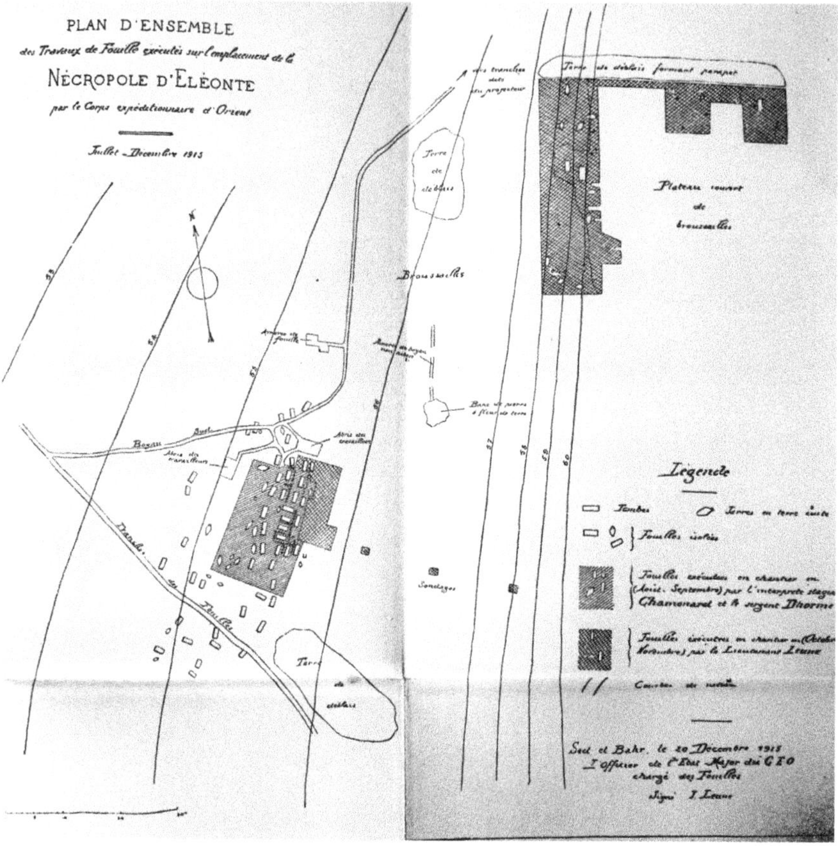

Figure 3.8 Plans of the Necropolis at Elaious.
Credit: Bulletin de Correspondance Hellénique XXXIX 1915

while the remaining 300 can be found in Istanbul's Archaeology Museum[23] (Figures 3.8 and 3.9).

Prior to the French excavations at Elaious, Ottoman laws forbidding the removal of antiquities from the empire had been proposed as early as February 1869.[24] The export of antiquities and archaeological remains from Ottoman territories was further restricted by the sultan's government with additional bylaws passed in 1874, 1884, and 1906.[25] This legislation enacted by the Ottoman government remained in force throughout World War One and the Occupation years (1918–23) after the war, so the removal of antiquities and unauthorized excavations of ancient sites was prohibited.[26] Nevertheless, archaeologists working for the governments of the former Allied countries took advantage of the disarray in Ottoman lands to forage for antiquities.[27] The French returned after the war during the period of foreign occupation and carried out more systematic archaeological excavations at Elaious and the "Tumulus of Protesilaus,"

Une tranchée des fouilles d'Eléante.

Figure 3.9 Archaeological finds from Elaious during World War One.

Archive L. Thys-Şenocak

Un soldat archéologue français sous sa tente, près de Seddul Bahr.

Figure 3.10 One of the French archaeologists seated with his excavation finds from Elaious.

Credit: Archive of L. Thys-Şenocak

the aforementioned Karaağaçtepe.[28] Military barracks which had been constructed inside the Ottoman fortress of Seddülbahir by the French during the war were repaired and refurbished, and used to house the archaeologists and their findings (Figure 3.10). At this time, graves of several of the French soldiers, who had died during the war and had been buried within the fortification, were also exhumed and their remains were to have been reburied in the cemetery close to ancient Elaious at Morto Bay. While most of the French soldiers who had died on the

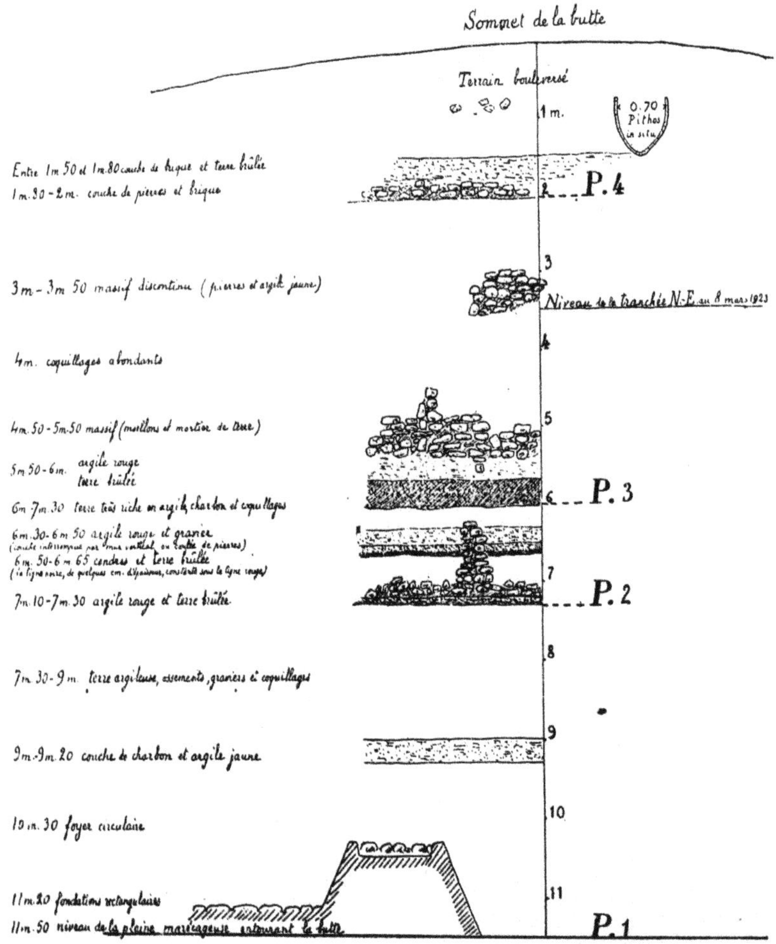

Fig. 11. — Stratigraphie schématique de la *toumba* de Protésilas (paroi Nord-Est).

Figure 3.11 Demangel's sketch of the stratigraphy of the Protosilaus mound.
Credit: Robert Demangel, *Le Tumulus Dit de Protésilas* (Paris: de Broccard, 1926)

peninsula were reburied in the French National Cemetery at Morto Bay, the 2017 excavations at the fortress of Seddülbahir revealed that several of their remains (including those of Théophile Marie Guyomard, who had died on 4 June 1915 at 34 years old, and was buried with his identity medallion), along with parts of the tombstones from the original French Cimetière Galinier graveyard in the fortress, had been forgotten by the post-war archaeologists.[29]

Located on the outskirts of the village of Seddülbahir, Karaağaçtepe is among the largest mounds on the peninsula, with eleven meters of archaeological levels.[30] Initially the French, like Schliemann before them, believed this to be a tumulus, but the French excavations revealed prehistoric levels corresponding to the phases of the levels of Troy I and II. Demangel's sketch of the stratigraphy of the site indicates the many levels of the mound; he noted that it appears to have been abandoned by the Middle Bronze Age and then reoccupied again in the Byzantine era[31](Figures 3.11 and 3.12). Traces of masonry architecture, some still visible today, were discovered, along with pottery, figurines, weights, and various stone tools, all from the earliest phase of the mound, which was found at a depth of 11.5 meters, and which Demangel believed could be dated to approximately 3000 BCE. Demangel noted that ceramics found at these early levels at Karaağaçtepe resembled pottery from similar eras at Troy, suggesting evidence for contact between these two shores of the Dardanelles.[32]

At this site, and throughout the Gallipoli peninsula, there are few material remains from the Middle and Late Bronze Age through the Iron Age, with the

Figure 3.12 Byzantine architectural remains discovered by Demangel.

Credit: Robert Demangel, *Le Tumulus Dit de Protésilas* (Paris: de Broccard, 1926)

exception of the site of Maydos/Kilisetepe. Habitation seems to have flourished again from the 9th century BCE through the 5th century BCE with colonization from the Greek mainland. An interest in the grain trade and the forests of the peninsula may have been the catalyst for the Greeks to begin their expansion as they had depleted their timber reserves closer to home.[33] Eventually the peninsula came firmly under the control of Athens and the Delian League by 466 BCE; Athens then struggled with Sparta during the fifth century BCE to maintain control over the lucrative grain trade that crossed the Black Sea and was channelled through the Hellespont.[34] In the last battle of the Peloponnesian Wars, Sparta was victorious on the Hellespont and secured its hold on the Gallipoli peninsula in a momentous battle in 405 BCE at Aegospotami in which thousands of Athenians lost their lives.[35] The fortunes and control of the peninsula continued to vacillate as it changed hands for the next three centuries among Athens, Sparta, the Persians, invading Thracian tribes, Macedonians, and eventually Alexander the Great.[36] The aforementioned dedicatory inscription to Attalus II, the kind of Pergamon (220–138 BCE) "savior and benefactor," was written at that time, only to be unearthed centuries later during the battles of World War One by Sergeant-Major R.S. Jones while stationed at his post in Suvla Bay.

Demangel notes in his archaeological excavation reports that despite the heavy military trenching in the Cape Helles region of the peninsula, no other traces of ancient sites apart from the sites of Elaious and Karaağactepe were found.[37] The latter tumulus has been surveyed in recent years by Reyhan Körpe from the 18 March University in Çanakkale; what was left after the French excavations of any stratigraphic integrity has been destroyed by farming and the illicit digging of treasure hunters. Archaeological excavations that could confirm, or at least revise, the findings of Demangel and his post-war excavations at Elaious are no longer possible as the Çanakkale Martyrs Memorial was constructed on this same site by the Turks in 1944.[38] There were no discussions concerning the findings of the earlier wartime excavations by the French, or any suggestions that scientific archaeological research should resume at that location, prior to the construction of such a large commemorative structure on the same site; or later in 2007 when a new symbolic cemetery for Çanakkale's martyrs was added to the Çanakkale Martyrs Memorial.[39]

Archaeology of another sort—the exhumation of the remains of the dead—began two years after the war had ended. In the summer of 1918 the British Graves Registration Unit (GRU), which had taken responsibility for the excavation, burial, and commemoration of the dead of World War One in the battlefields and war cemeteries throughout Europe, decided that there was a crucial need to identify the remains of Allied troops on the peninsula and rebury them in a more orderly fashion. The British government, in pragmatic efforts to keep commemorative activities fair and less costly, had decided against the repatriation of the remains of their war dead in a process that Thomas Lacquer has referred to as "a uniformity of memorialization."[40]

Excavation by the GRU was driven by the need of the Commonwealth governments to ascertain the identity of a fallen soldier for commemorative purposes and to make an account of the death for state records and the families of the dead.[41] Thomas Pemberton, who worked after World War One for the CWGC on the peninsula (the organization that succeeded the GRU and still manages the ANZAC and British graves there), recounted the methods of identification used for these types of "surface findings" and how they were extracted from the earth. In 1919 he reported quite clinically:

> Three years had been more than enough to reduce all human remains to clean whitened bones, for where the earth teems with insect life under the hot summer sun organic matter is subject to rapid change. But pieces of uniform, metal badges, buttons, knives, equipment were well preserved. Indelible ink on oil sheets was often found to be as clear as on the day it was written. Much time and industry were devoted to searching for identity disks. If identification was not immediately established the ground in the vicinity of human remains was sieved and searched."[42]

The reports about these graves detected by former Allied military-personnel-turned-forensic-archaeologists reveal that the team sent to exhume bodies quickly learned to detect signs of recent burials. One member of the team noted:

> Where a grave had been dug the loosened soil lost its moisture more rapidly than the set and hardened earth, and the herbage above dried up slightly in advance of the surrounding growth. Wherever this differentiation in the condition of the grass was observed there was almost sure to be a grave underneath."[43]

Collecting "trophies" and "national relics"

By the winter of 1919 the majority of the excavations of human remains by the GRU had been completed and work now turned to the planning of cemeteries, commemorative monuments, and the creation of national museums back home. Australians recognized very soon after the war the value of wartime objects left on the peninsula as part of their national patrimony, and Charles E. Bean, the official Australian war historian, set about retrieving a massive collection of wartime objects that had been left on the peninsula during the evacuation, several of which he described as "national relics" or "trophies."[44] These were put on display in different museums in Australia, including the Australian War Museum in Melbourne between 1922 and 1925, and attracted over one million visitors during those years (Figure 3.13).[45] In several messages to the GRU, there are directives from Bean to salvage items that had been left on the battlefields and along the shores of the Gallipoli peninsula during the 1916 evacuation because it was believed they would be of interest to museums back in Australia; alternatively

Figure 3.13 The interior of the Australian War Museum in Melbourne on or about the day of its opening, 25 April 1922.

Credit: AWM J00290

he suggested they could be sold to raise funds for such a museum. Internal correspondence between Bean's team and the Australian headquarters in Constantinople describes the material they had or hoped to salvage from the beaches of Gallipoli, which had been evacuated three years earlier. The ordnance and wartime memorabilia were valued as collectibles, and efforts were made to get hold of these before the Turks could. Many wartime objects were vulnerable and Bean noted in a hastily composed telegram that:

> . . . if unremoved [they] will [become] Turkish war trophy. After Anzac graves path completed these should make quixkest comprehensivest salvage [of the] whole area [in] order [to] remove national relics.[46]

Large items such as boats, and even wooden support posts from bunkers and dugouts, were considered to be collectible, and there was an urgency to get these off the peninsula before the locals could retrieve them. Turks living on the peninsula had also been salvaging wartime ordnance and memorabilia but were using these mainly to rebuild homes and eke out a living from the scrap value. Wartime finds were not regarded as museum quality artifacts by either the locals or the officials representing the Ottoman government. For the Australians, however, the panic to salvage as much as possible for future exhibition or to sell in Australia, and to leave as little as possible behind, is evident in the many telegrams sent by Bean to the home office.

> There are still a number of trophies at Anzac which could well be collected; two ships lifeboats on the beach; the barrel of our old 4.7-in gun, left by the roadside near several other trophies, besides any amount of the smaller stuff—bullets, loophole plates, shell fragments, timbers taken from Lone Pine and now found in stables and huts built by the Turks at the back of the Pine. I am suggesting in reference to these that Lieut. Hughes be given a party of 100 Egyptian Laborers, and a small Australian staff chosen by himself, and after fixing up the graves and paths be authorized to make a quick but comprehensive salvage of the whole area. He would label the stuff by the place in which he found it; and if he collected too much for use for our Museums, that the balance be used for exchange with other Museums, or for the sale to private visitors at our Museums for the benefit of the museum fund. . . . Hughes is a capable chap and could take this on for a month after finishing his graves. If he doesn't, the natives will clear the place quickly.[47]

Archaeological surveys and excavations on the Gallipoli peninsula since 1979

After a hiatus of almost sixty years, archaeological interest in the Gallipoli peninsula began again at different sections of the Gallipoli peninsula with four non-invasive archaeological survey projects conducted in 1979, 1989, 1997, and 2003.[48] Archaeological excavation began again on the peninsula in 2005 at the site of the Ottoman fortress of Seddülbahir with a joint team from Koç University, Istanbul Technical University, and the Çanakkale Archaeology Museum. Sixty-five years after the French excavations of the "Protesilaus tomb" by Demangel had come to an end, the excavations in the Cape Helles section of the peninsula at Seddülbahir, classified by the Turkish Ministry of Culture and Tourism as a salvage (*kurtarma*) excavation because the archaeological work preceeded a conservation project, marked a new phase in Gallipoli's history.[49]

As indicated in Table 3.1, archaeological work in Turkey is defined, divided, and excavation permission granted by the government according to several categories.[50] Research by archaeologists at a specific site, or in a designated region of Turkey, can be classified as a non-invasive archaeological survey of surface finds, or as an archaeological excavation if more invasive methods are used. Classification also includes the purpose of the survey or excavations and whether the goals of the work are for academic research, surface cleaning, salvage digging for roadworks, dams, or other infrastructure projects such as the Trans-national Anatolian Natural Gas Pipeline Project. Categories also exist to differentiate whether the excavation or survey was conducted by the local museum, and whether the director of the survey or excavation is a Turk or foreigner.[51]

Table 3.1 Archaeological excavations and surveys on the Gallipoli peninsula, 2006–16

Year	*Locations of surveys*[1] *and excavations on the Gallipoli peninsula from 2006–2016*	*Director/Co-director*	*Type of survey/excavation*
2006	Seddülbahir Fortress at Cape Helles	Çanakkale Museum, Koç University and Istanbul Technical University	Museum Salvage Excavation (Late Ottoman, Turkish and World War One periods)
	Gelibolu (Bolayır, Bakla Manastır mevkii/Eceabat – Büyük Anafartalar – Küçük Kemikli)	Mustafa Sayar	Field Survey (Epigraphy)
	Gallipoli National Historic Park and peninsula	Onur Özbek	Field Survey (Prehistory)
	Gallipoli National Historic Park and peninsula	Osman Uysal	Field Survey (Medieval and Turkish Periods)
	Aegean and Mediterranean	Harun Özdaş	Underwater Archaeology Survey
2007	Gallipoli National Historic Park and peninsula (Akbaş Şehitliği, Asartepe, Bahçeköy, Bağtepe (Höyük), Beşyol, Hamaylıtarla (Höyük), Çakıllık, Değirmenlik (Höyük), Hamamtepe, Kalanuro (Höyük), Kaynarca (Höyük), Keçili Mezarlığı, Güneyli Limanı (Höyük), Örenler Mevkii, Lale Tepe, Tekke Tepe; Demirtepe Köyü, Ece Limanı, Sütlüce Köyü, Münipbey Deresi)	Onur Özbek	Field Survey (Prehistory)
	Gallipoli National Historic Park: (Eceabat, Keçili, Yalova, Bahçeköy, Büyük Anafartalar, Küçük Anafartalar, Kumköy, Bigalı Kalesi)	Osman Uysal	Field Survey (Medieval and Turkish Period)
	Çanakkale, İzmir, Muğla, Antalya, Mersin	Harun Özdaş	Underwater Archaeology Survey
2008	Mecidiye Fortress (Kilitbahir)	Çanakkale Museum	Museum Salvage Excavation
	Gallipoli National Historic Park and peninsula (Tuz Gölü havzası, Küçük Anafartalar, Büyük Anafartalar, Büyük Kemik Burnu Doğusu, Ece Limanı Batısı, Asma Dere, Değirmen Dere, Çakal Dere, Eceabat Batısı: Kocadam Mevkii, Eğerli Tepe, Çınarlı Dere, Kakma dağ, Geren Ovası,	Onur Özbek	Field Survey (Prehistory)

	Azmak Dere, Eğerli Tepe, Bigalı Köyü, Hacı Hüsrev Höyük, Kilye Ovası, Havuzlar Mahallesi/Mevkii, Gözetleme Tepe/Kel Tepe Höyük, Karaağaçtepe, Behramlı Köyü)		
	Gallipoli National Historic Park and peninsula (Bolayır, Maltepe Mevkii, Çokal Köyü, Karainebeyli'de Hisarlık, Kiliselik, Beşyol, Ece Limanı Ilgardere, Musal Tepe, Küçük Kemikli, Cumalı, Eski Burhanlı, Süleymaniye Köyü)	Osman Uysal	Field Survey (Medieval and Turkish Periods)
	Çanakkale, İzmir	Harun Özdaş	Underwater Archaeology Survey
2009	Mecidiye Fortress, Kilitbahir	Çanakkale Museum	Museum Salvage Excavation
	Gallipoli National Historic Park and peninsula	Onur Özbek	Field Survey (Prehistory)
	Gallipoli National Historic Park and peninsula	Osman Uysal	Field Survey (Medieval and Turkish Periods)
	Çanakkale, İzmir, Muğla, Antalya, Mersin	Harun Özdaş	Underwater Archaeology Survey
2010	Maydos Kilisetepe Eceabat	Göksel Sazcı	Excavation
	Gallipoli National Historic Park and peninsula	Onur Özbek	Field Survey (Prehistory)
	Gallipoli National Historic Park and peninsula	Osman Uysal	Field Survey (Medieval and Turkish Periods)
	Gallipoli National Historic Park and peninsula[2]	Ayşe Çaylak Türker	Field Survey (Early Christian and Byzantine Period)
	Arıburnu (Anzac areas)	Mithat Atabay/Antonio Sagona	Field Survey (Çanakkale Wars)
	Çanakkale, İzmir, Muğla, Antalya, Mersin	Harun Özdaş	Underwater Archaeology Survey
2011	Maydos Kilisetepe Eceabat	Göksel Sazcı	Excavation
	Gallipoli National Historic Park and peninsula	Onur Özbek	Field Survey (Prehistory)
	Gallipoli National Historic Park and peninsula	Osman Uysal	Field Survey (Medieval and Turkish Period)

continued . . .

Table 3.1 Continued

Year	*Locations of surveys*[1] *and excavations on the Gallipoli peninsula from 2006–2016*	*Director/Co-director*	*Type of survey/excavation*
	Gallipoli National Historic Park and peninsula	Ayşe Çaylak Türker	Field Survey (Early Christian and Byzantine Period)
	Arıburnu (Anzac areas)	Mithat Atabay/Antonio Sagona	Field Survey (Çanakkale Wars)
	Gelibolu, Eceabat, Büyük Anafartalar Mezarlığı, Küçük Kemik Burnu, Sestos/Akbaş limanı, Cumalı Köyü, Bolayır, Bakla Burnu, Maltepe, Tayfur Köyü	Mustafa Sayar	Field Survey (Epigraphy)
2012	Maydos Kilisetepe Eceabat	Göksel Sazcı	Excavation
	Gallipoli National Historic Park and peninsula	Osman Uysal	Field Survey (Medieval and Turkish Period)
	Gallipoli National Historic Park and peninsula	Ayşe Çaylak Türker	Field Survey (Early Christian and Byzantine Period)
	Arıburnu (Anzac areas)	Mithat Atabay/Antonio Sagona	Field Survey (Çanakkale Wars)
		Mustafa Sayar	Field Survey (Epigraphy)
2013	Maydos Kilisetepe Eceabat	Göksel Sazcı	Excavation
	Gallipoli National Historic Park and peninsula	Osman Uysal	Field Survey (Medieval and Turkish Period)
	Gallipoli National Historic Park and peninsula	Ayşe Çaylak Türker	Field Survey (Early Christian and Byzantine Period)
	Arıburnu (Anzac areas)	Mithat Atabay/Antonio Sagona	Field Survey (Çanakkale Wars)
	Çanakkale, Muğla, Antalya	Harun Özdaş	Underwater Survey
2014	Maydos Kilisetepe Eceabat	Göksel Sazcı	Excavation

	Gallipoli National Historic Park and peninsula	Onur Özbek	Field Survey (Prehistory)
	Gallipoli National Historic Park and peninsula	Ayşe Çaylak Türker	Field Survey (Early Christian and Byzantine Period)
	Arıburnu (Anzac areas)	Mithat Atabay/Antonio Sagona	Field Survey (Çanakkale Wars)
	Gallipoli peninsula, Sestos	Reyhan Körpe	Field Survey (Classical)
	Çanakkale, Muğla, Antalya	Harun Özdaş	Underwater Archaeology Survey
2015	Maydos Kilisetepe Eceabat	Göksel Sazcı	Excavation
	Gallipoli National Historic Park and peninsula	Osman Uysal	Field Survey (Turkish Period)
	Gallipoli National Historic Park and peninsula	Ayşe Çaylak Türker	Field Survey (Early Christian and Byzantine Periods)
	Gallipoli National Historic Park and peninsula (Kilye, Poyraz Tepe, Bigalı, Ece Limanı, Büyük Kemikli Burnu, Kireçtepe Batısı, Anafartalar Ovası, Büyük Anafartalar Köyü, Çamlıtekke Mevkii, Seddülbahir Köyü)	Reyhan Körpe	Field Survey (Classical)
2016	Maydos Kilisetepe Eceabat	Göksel Sazcı	Excavation
	Gallipoli National Historic Park	Osman Uysal	Field Survey (Turkish Period)
	Gallipoli National Historic Park	Ayşe Çaylak Türker	Field Survey (Early Christian and Byzantine Period)
	Gallipoli peninsula, Sestos	Reyhan Körpe	Field Survey (Classical)

1 Specific locations of surveys are given in Turkish and when that information was available
2 Data for 2017 appeared as this book was in press so it was too late to include in the table here but it is accessible at http://www.kulturvarliklari.gov.tr/TR,200465/2017-yili-kazi-ve-yuzey-arastirma-faaliyetleri.html

While Table 3.1 shows all survey and archaeological campaigns undertaken on the Gallipoli peninsula from 2006 through 2016, data for archaeological surveys and excavations conducted for academic research purposes prior to 2006 can be found in the annual field reports written by the survey or excavation director and published by the Turkish Ministry of Culture and Tourism. These earlier reports and related publications show that a series of non-invasive surveys on the peninsula by Mehmet Özdoğan of Istanbul University were initiated in 1979 and concluded in 1982 as part of his larger research project about the Prehistoric sites of Northwestern Turkey.[52] At that time a total of 57 ancient sites were documented; of these, 14 had Prehistoric remains with the majority revealing Bronze Age and Early Bronze Age material.[53] Özdoğan noted then that the mound of Kilisetepe in today's Eceabat was likely the most substantial settlement on the peninsula during the Early Bronze Age and pointed out that there were connections to corresponding levels at Troy. During this first post-war survey of 1979 the locations of several Classical and Medieval sites which had previously been known only from texts were documented.[54] From 1989 through 2007, Mustafa Sayar of Istanbul University conducted epigraphic surveys in eastern Thrace, including the Gallipoli peninsula. Among the many finds Sayar noted during his work on the Gallipoli peninsula were the discovery of a Hellenistic era statue from Bolayır, now in the Archaeology Museum of Çanakkale, and Latin and Greek inscriptions from spolia at several different locations on the peninsula.[55]

The most comprehensive survey and inventory of the natural and cultural assets on the Gallipoli peninsula were undertaken in 1997 by Raci Bademli of Middle East Technical University (METU). After a major fire devastated 4049 hectares of the peninsula's forest land in 1994, Bademli was invited by the government to initiate an inventory of the natural and cultural assets of the peninsula survey; this resulted in a formidable two-volume publication entitled the Long Term Development Project (UDGP). The preliminary inventory of the survey would be used as a resource for the Gallipoli Peace Park International Ideas and Competition, held in 1997–98; thereafter it served for over a decade as the management plan for the Ministry of Forestry and Water Affairs for the assets of the peninsula.[56] Most importantly, Raci Bademli and his team proposed in the UDGP that the peninsula become an open-air museum.[57]

When the 1997 report by Bademli was compiled there were only four small museums on the peninsula, including a house museum and an open-air museum that had official museum status. All these state-owned museums displayed objects and material culture from World War One. In addition to these museums, many World War One artifacts and memorabilia were collected and displayed in private collections throughout the peninsula, called "war galleries." These collections, while not officially designated as museums, were legally allowed to charge visitors and exhibit material culture from World War One. The majority of the archaeological objects found on the peninsula, along with those dating from the war (particularly several larger artifacts such as cannons), were stored or displayed in larger state and/or military museums either in Çanakkale or Istanbul, or just left in situ exposed to the elements.

While the extent of the documentation of natural and cultural heritage assets and sites on the peninsula was admirable, the UDGP did not survey the World War One trenches in a detailed manner.[58] The need for that type of documentation was recognized by a team from Australia which undertook in 2003 a four-day survey of the "Old Anzac area," which included the ANZAC battlefields on the Second Ridge of the peninsula. The directors of that survey noted at that time "the need for cultural heritage practitioners to be involved in the management of the sites within the Gallipoli Peninsula Peace Park."[59] This initial Australian-led survey helped to lay the groundwork for a more comprehensive five-year non-invasive archaeological survey and documentation project conducted by the Australian National University in Melbourne and the 18 March University in Çanakkale: the Joint Historical and Archaeological Survey (JHAS).

The fieldwork for the non-invasive survey was limited to the ANZAC section of the peninsula and ran for four years between 2010 and 2014; it was directed by Antonio Sagona from the Australian National University in Melbourne and Mithat Atabay from 18 March University in Çanakkale.[60] The wartime artifacts that the JHAS found in the ANZAC area were deposited, after consultation with the Turkish Ministry of Culture and Tourism, in the Çanakkale Naval Museum (Çanakkale Deniz Müzesi).[61] Smaller finds of human remains, such as bone fragments, were dealt with according to the guidelines stipulated for these types of remains by the Office of Australian War Graves in the Department of Veterans' Affairs, which advocates reburying the bone fragments back into the earth of the battlefield.[62] Because the team could not often determine the nationality of the human remains discovered, when these findings were encountered "the normal practice was for a team member from Turkey and one from the Anzac contingent to rebury the fragment and then show respects in an appropriate manner."[63] As C.E. Bean had noted a century earlier in his efforts to identify and rebury human remains of the Australian dead on the Gallipoli landscape, in many cases the national origins of the bones were difficult to identify as Australians and Turks were frequently found buried together.[64]

Surveys of the Prehistoric-era remains on the peninsula, conducted by Onur Özbek of 18 March University in Çanakkale between 2006 and 2008, have revealed important Early Neolithic finds in the region, such as red slipped and burnished ware at the site of Hamaylıtarla at the far northern entrance to the Gallipoli peninsula. Evidence from this site, along with Kaynarca and Karaağaçtepe at the southern end of the peninsula, attests to the importance of the region in these early Prehistoric eras.[65] Ayşe Türker from 18 March University was granted permission by the Turkish Ministry of Culture and Tourism in 2010 to document the much later phases of the peninsula's past; here Türker found early Christian and Byzantine remains while surveying both sides of the Dardanelles, including the towns of Eceabat and Gallipoli.[66] The archaeological excavation of the Kilisetepe Settlement Mound north of Eceabat and the excavation at Çamburnu at the southern end of Eceabat began in 2010; both were supervised by archaeologists from 18 March University.[67] Levels dating from the 3rd millennium BCE to the early twentieth century confirmed Özdoğan's earlier observations made in his survey of much earlier periods of occupation and attest to a continued usage of the site of the Maydos Kilisetepe Settlement Mound as a base for seafaring trade and strategic military control of the straits.[68] The excavation, initiated on the peninsula by the Çanakkale Archaeology Museum and the 18 March University was led by Göksel Sazcı and has already revealed a rich array of architectural remains, ceramics, and other archaeological material from the Bronze and Iron Ages, Archaic Greek and later Byzantine and Ottoman eras. Burials at least a century old were discovered in the Hagios Dimitrios Church on the site. A center for wine production, which may date to the Byzantine

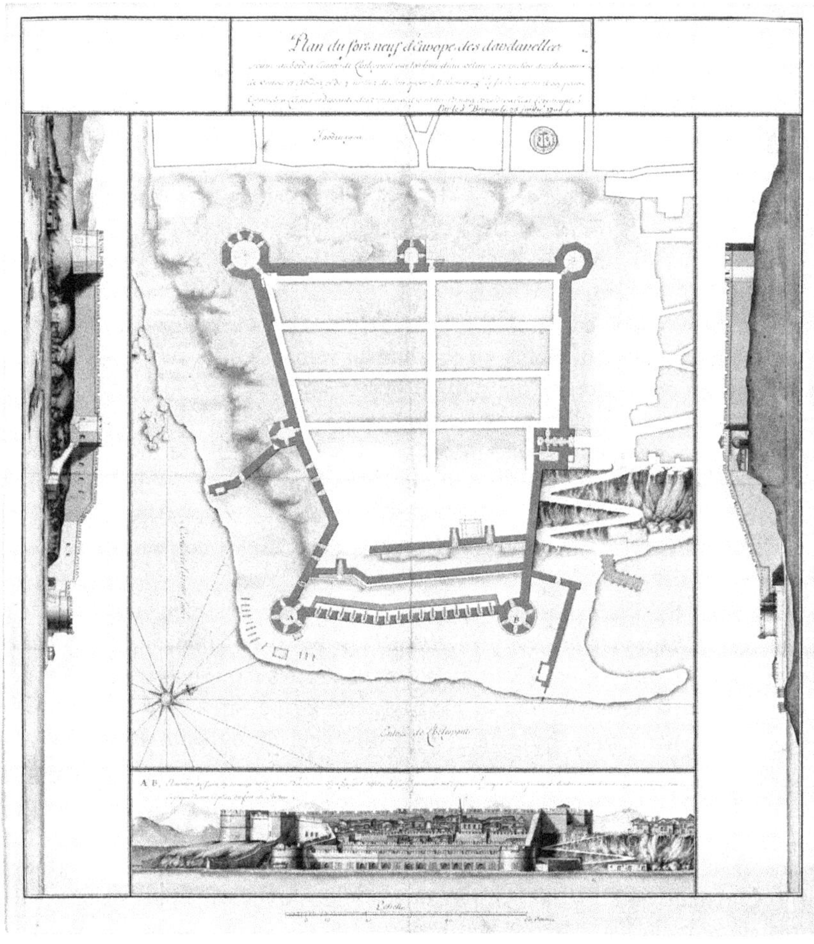

Figure 3.14 The fortress of Seddülbahir drawn by Berquin in 1700.
Credit: BnF Cartes et Plans, G.F.98/4/2

occupation of the site from the thirteenth to the fifteenth century, was also discovered in nearby Çamburnu by the same team.[69]

The excavations and documentation done at Seddülbahir in 2005–6 and 2016–18 have been instrumental in establishing Ottoman archaeology and the archaeology of World War One as important archaeological subfields in Turkey. These excavations of later eras point to the role these previously neglected levels of Turkey's landscape can have in helping to achieve a greater understanding of the region's more recent past. The Seddülbahir fortress project contributed to the development of a methodology for the excavation and restoration of that particular type of Ottoman structure—the fortification—by using archival records to inform decisions about where to excavate and how to conserve an important monument of the Ottoman frontier (Figure 3.14). Situated at the entrance to the Dardanelles, Seddülbahir was built in the mid-seventeenth century by Hadice Turhan Sultan, the mother of Sultan Mehmed IV, to prevent the Venetians and pirates from passing through the straits and besieging Istanbul.[70] In the archaeological excavation that was conducted there from 2005 to 2006, both European and Ottoman archival records were used to understand 300 years of repairs and modifications that were made to the fortress, and to determine where excavations should be conducted. Using the rich textual and visual records together with an

Figure 3.15 Seddülbahir fortress showing the destruction of the west tower during World War One.

Credit: Imperial War Museum, Q13233

assessment of the physical ruins and material remains at the site revealed that while there had been several instances of remodeling the fortress, the majority of the changes that had occurred over its three centuries of existence were the result of successive earthquakes or bombardments to the site during World War One. The battle on 25–26 April 1915 on the V beach just below the fortress was among the most destructive in the entire campaign and took a terrific toll on the structure. The decision of the restoration team working with ÇATAB at Seddülbahir to leave the west and south towers of the fortress as ruins thereby preserving tangible evidence of the destruction of World War One insures that there will continue to be evidence, and lieux de mémoire, of the violence and destruction that were a crucial part of the history and heritage of the peninsula during World War One (Figure 3.15).[71] In an effort to bring together the archaeological and historical understanding of the site and shed light on the concerns of the present community in Seddülbahir, an oral history project was also conducted at Seddülbahir from 1999 to 2002. This provided much-needed insight into the more recent past of residents whose families had been living in the environs of the fortress since the 1930s.[72] In this palimpsest of Gallipoli's landscape, the Ottoman layers need to be properly documented, along with the more recent archaeological past of World War One, and each era given the same kind of attention and respect by archaeologists as earlier periods of occupation.[73]

The most recent archaeological surveys and excavations on the peninsula were initiated in 2015 by archaeologist Reyhan Körpe from 18 March University in Çanakkale, who also participated in the JHAS project in the ANZAC region. The fieldwork that Körpe conducted during the 2016 season included ten areas in several sections of the Gallipoli peninsula.[74] Mert Çatalbaş, the lead archaeologist of ÇATAB, has supervised the salvage archaeological excavations at the fortresses of Seddülbahir and Biga from 2015 to the present.[75]

New technologies, new methodologies for invasive and non-invasive archaeology

Many years of military restrictions on surveys and archaeological excavations of the Gallipoli peninsula did limit the extent of the archaeological data available for this important region which links Thrace to Asia. However, one benefit of this hiatus was that once research was finally allowed many projects used state-of-the-art invasive and non-invasive survey and archaeological technologies, and employed new methodologies to shed light on a range of questions surrounding the settlement patterns of peoples residing on the peninsula in different eras, from the Neolithic through the Byzantine, Ottoman, and World War One periods.

New technologies for land-based survey and archaeological excavations were used on the Gallipoli peninsula as early as 2005, when the documentation project of the Seddülbahir fortress conducted the first comprehensive 3D laser scanning of a historical monument in Turkey. Survey data for that site, which had been collected during the first phase of that project in 1997, was integrated as early as 2000 into the project's Geographical Information System (GIS).[76] In 2010, following the Seddülbahir survey, the JHAS non-invasive archaeological survey and documentation project began its explorations of the ANZAC landscape, using new non-invasive techniques to research World War One landscapes, trenches, cemeteries, earthworks, and material remains.[77] The team used traditional methods of field walking, but also employed remote sensing methods such as ground-penetrating radar (GPR) and analyses of aerial and satellite images to understand the spatial dimensions of human occupation and wartime conflict in the 4.2 km^2 area they investigated. The team also established a GIS called the ANZAC Gallipoli Archaeological Database (AGAD) in order to "assist the study of the First World War through its emphasis on landscape and artifacts."[78]

The JHAS team worked at a selection of sites in the ANZAC area; some of these included WWI era cemeteries such as Lone Pine and Johnston's Jolly.[79] Surface finds along with topographical features, which included 16.5 km of trenches and 82 tunnels, were documented with differential global positioning system (DGPS) technology using a hand-held global positioning system (GPS). All measurements and data were also integrated into the project's GIS digital database. Satellite imagery, historical Ottoman and Allied maps, and photographs were brought together in the GIS to better understand the relationship of the topography to the material culture the team discovered on this part of the peninsula's landscape, and to investigate how the landscape had changed since World War One by examining the "relationships between the battlefield landscape of 1915 and the tracks, fire trails, roads and memorials that form the commemorative landscape today."[80]

Among the most recent projects to use non-invasive archaeological methods to explore the World War One past of the peninsula is the geophysical survey of the Ottoman cemetery of Ağadere just outside the village of Kilitbahir, where approximately 3,000 soldiers were buried adjacent to the wartime hospital that was once there. This research, supervised by the Departments of Geophysical and Geomatics Engineering at 18 March University in Çanakkale, was based on the aforementioned Şevki Pasha maps of the region which were completed by the Ottomans shortly after the war ended and indicated various features of the peninsula including trench systems and mass burial areas.[81] The team employed a range of non-invasive geophysical methods of investigation, including GPR, electrical resistivity tomography (ERT), and magnetic imaging (MI), because of what they perceived as the ethical and religious implications of using more invasive methods at an Ottoman cemetery. Among the team's explanations for their non-invasive approach to the site was that "excavation is impossible in Turkey in view of the respect given to martyrdom."[82] While sensitivity may have been demonstrated in the Ağadere project by using a non-invasive approach to archaeological excavation, the project for which this research was conducted the Ağadere Hospital and Memorial complex – is a massive construction which was completed in 2015 at the site and has still not opened. Some recent commemorative projects on the peninsula have had no preliminary scientific archaeological research conducted. This was the case for the 2012 Legend of Gallipoli Simulation and

Information Center at Kabatepe, the Akbaş Memorial complex and the 2018 Alçıtepe Mosque complex. In these cases foundation work and construction was undertaken without archaeological documentation in spite of the likelihood of these sites having human and other remains from World War One and earlier eras. These projects are discussed in greater detail in the final chapter.

Underwater archaeology and cultural heritage

Among the important developments in archaeology and cultural heritage management of the peninsula is that the research about the Gallipoli peninsula has pushed beyond its land borders to encompass the surrounding maritime areas and the shipwrecks in the waters of the Dardanelles and the Marmara and Aegean seas. Since the 2014 centenary of World War One, the underwater heritage of the Gallipoli peninsula, including all shipwrecks that have been in the sea for a century or longer, could come under the protection of UNESCO's 2001 Convention on the Protection of the Underwater Cultural Heritage.[83] Unfortunately, the salvaging and selling of many of the wrecks was permitted by the Turkish Ministry of Defense in 1923, and rights to export the scrap metal from the underwater wrecks were given to a series of foreign companies so that the materials could be used for national needs.[84] Successive government bans on salvaging the historic wrecks over the next five decades were driven more by the need to keep valuable scrap metals such as copper and copper alloys in the country rather than an interest in protecting the remains of historic shipwrecks, as many Turkish companies were allowed to harvest metal and other materials from these wrecks. Shipwrecks in the Dardanelles were particularly compromised by salvage operations from 1957 to 1970. After 1970, as Selçuk Kolay remarked, "the salvage operations were no longer economically viable. In addition to this, there were concerns about losing the wrecks, considered to be historical sites and the protection of war graves. These two issues were the reasons behind the decision to terminate further salvage operations."[85] Surveys of underwater cultural assets from World War One have been conducted in Turkey both in academic and professional salvaging projects since 1997, and permission has been granted by the Turkish Ministry of Culture and Tourism for underwater surveys and excavations in the Çanakkale region since 2007.[86] However, Turkey has yet to ratify the UNESCO 2001 Convention on the Protection of Underwater Cultural Heritage and use this instrument for managing its underwater heritage from World War One (and earlier eras) that exists off the coasts of the Dardanelles and the Gallipoli peninsula.[87]

Underwater archaeology of the battlescapes off the coast of the Gallipoli peninsula is still in an early phase, but new technologies such as 3D multibeam sonar imaging have helped to document more than 30 known wrecks of submarines, gunboats, and other elements of World War One maritime heritage (including mines) that remain in the waters of the region, many beyond the access of the individual diver (Table 3.2 and Figure 3.16).[88] While there have been several projects since 1998 to document these World War One

Table 3.2 Wrecks in the Dardanelles and coastal waters surrounding the Gallipoli peninsula[1]

Number on map	*Name of ship*	*Year launched*	*Nationality*	*Type*	*Sinking date*	*Wreck location*	*Depth in meters*
1	Mesudiye	1874	Ottoman	Battleship	13/12/1914	Dardanelles Straits	12
2	Saphir	1910	French	Submarine	15/01/1915	Dardanelles Straits	55
3	Bouvet	1896	French	Battleship	18/03/1915	Dardanelles Straits	72
4	Irresistible	1901	British	Battleship	18/03/1915	Dardanelles Straits	62
5	Ocean	1898	British	Battleship	18/03/1915	Dardanelles Straits	68
6	E15	1914	British	Submarine	15/04/1915	Dardanelles Straits	3
8	Joule	1912	French	Submarine	01/05/1915	Dardanelles Straits	44
10	Goliath	1898	British	Battleship	13/05/1915	Dardanelles Straits	74
11	Triumph	1903	British	Battleship	25/05/1915	Kabatepe	72
12	Majestic	1895	British	Battleship	27/05/1915	Seddülbahir	28
13	Carthage	1910	French	Passenger ship	04/07/1915	Seddülbahir	84
14	Mariotte	1912	French	Submarine	27/07/1915	Dardanelles Straits	3
15	Barbaros Hayrettin	1891	Ottoman	Battleship	08/08/1915	Marmara Sea	16
16	Lundy	1908	British	Minesweeper	16/08/1915	Suvla Harbour	28
17	E7	1913	British	Submarine	04/09/1915	Dardanelles Straits	42
18	Milo	1892	?	Cargo ship	26/10/1915	Arıburnu	7
19	Louis	1913	British	Destroyer	31/10/1915	Suvla Harbour	14
23	Midilli	1911	Ottoman	Cruiser	20/01/1918	Aegean Sea	74
24	E14	1914	British	Submarine	28/01/1918	Dardanelles Straits	20
25	Renarro	1913	British	Minesweeper	10/11/1918	Dardanelles Straits	70
26	Barge	?	Australian	Landing craft	?	Arıburnu	54
27	Barge	?	Australian	Landing craft	?	Küçük kemikli	30
28	Barge	?	Australian	Landing craft	?	Arıburnu	28
29	Mule Barge	?	Australian	Landing craft	?	Arıburnu	18
30	The Helles barges	?	British	Landing craft	?	Seddülbahir	26
31	The unknown wreck	?	?	Minesweeper	?	Alçıtepe/Sarıtepe	58
32	Atılay (suppl. wreck I)	1939	Turkish	Submarine	14/07/1942	Aegean Sea	68
33	Dumlupınar	1944	Turkish	Submarine	04/04/1953	Dardanelles Straits	84

1 This table is modified from S. Kolay et al, *Echoes from the Deep*, 58–59 and does not include the wrecks located in the Marmara Sea. The number of the wreck corresponds to the numbers on the map (Figure 3.16) provided by S. Kolay et. al, *Echoes from the Deep*, 62.

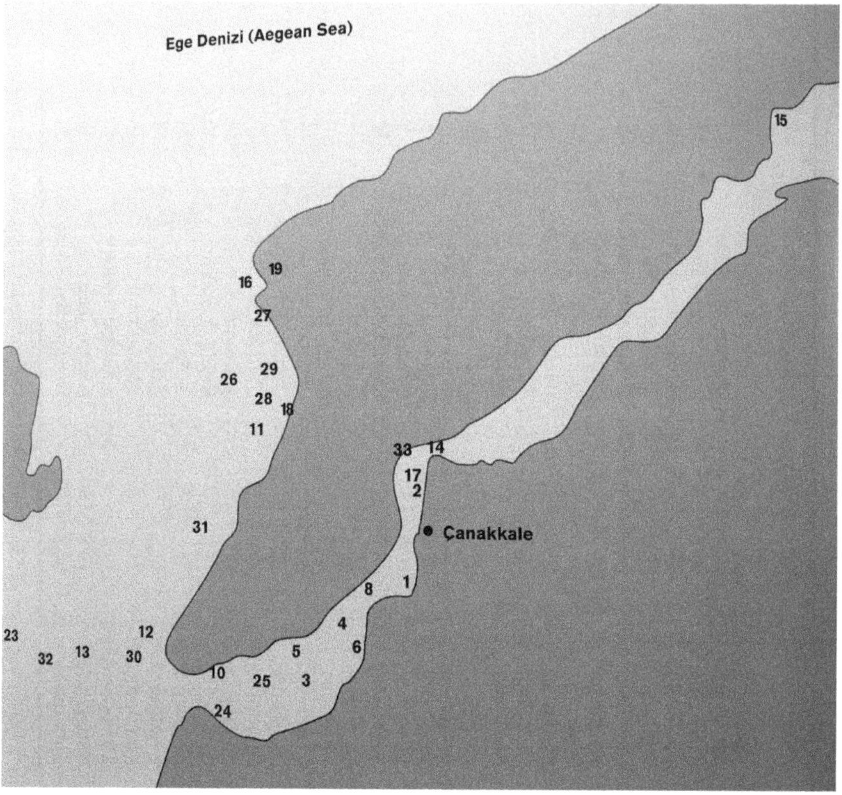

Figure 3.16 Map of the Gallipoli peninsula and the Dardanelles indicating the location of shipwrecks.

Credit: S. Kolay, O. Taktak, S. Karataş, M. Atabay, *Echoes from the Deep: Wrecks of the Dardanelles Campaign*; Vehbi Koç Foundation, 2013

wrecks and bring public attention to their heritage value, opinions differ among the teams involved as to whether these wrecks should be salvaged or left in situ.[89]

The first documentary film about the Dardanelles wrecks, *History in the Depths*, was made by Selçuk Kolay and Savaş Karakaş in 1998 and a second film, *Gallipoli: War Beneath the Waves*, in 2006.[90] Underwater surveys and documentation campaigns of the World War One wrecks in the waters off the shores of the peninsula have also been the focus of private companies such as Kolay Marine Ltd and Derinsu Ltd, which between 1998 and 2014 documented over 33 wrecks in the Aegean Sea, the Dardanelles, and the Sea of Marmara, including the first-time documentation of a French passenger ship, *Carthage*, a British submarine, *E14*, and a minesweeper, *Renarro*.[91] A Polish diving team led by Piotr Nykiel, Piotr Wytykowski, and Roman Zajder, the founders and co-chairmen of the Shipwreck Expeditions Society, documented six wrecks off the shores of the peninsula in 2011 using more traditional methods of documentation.[92]

New technologies and recommendations for the management and conservation of the underwater World War One wrecks in Gallipoli waters were proposed in a 2010 project entitled "Project Beneath Gallipoli," which hoped to examine submerged archaeological material at several of the main landing beaches surrounding ANZAC cove.[93] This research was conducted by a joint Australian–Turkish team led by the maritime archaeologist Tim Smith, who undertook a detailed sidescan sonar survey of the seabed around Brighton Beach, ANZAC cove, North Beach, and Suvla Bay.[94] ÇATAB has also recommended in its strategic plan for 2016–2020 that what remains of these important examples of Gallipoli's underwater heritage be protected and preserved.[95]

Battlefield archaeology, conflict archaeology, World War One archaeology

New methodologies and the emergence of subfields in archaeology have been as important as high-tech tools in efforts to explore and better understand the heritage of the Gallipoli region. These subfields of archaeology have brought innovative ways of conceptualizing research projects on the peninsula, interpreting the material culture discovered there, and managing the peninsula's varied cultural and natural resources. The JHAS project's non-invasive survey of the ANZAC region, mentioned above, employed new technologies for surveying and landscape analyses, but equally as important, this interdisciplinary team introduced new methodologies to investigate the landscape—developing theoretical frameworks for specific sub-disciplines of archaeology such as "battlefield archaeology" and the "archaeology of conflict." As Birkett-Rees has pointed out, "archaeology of conflict" as a subfield of archaeology began almost three decades ago with the study of historic and ancient battlefields, but the application of these new methodologies to the battlefields of World War One and World War Two has encouraged approaches which interrogate the nature of the conflict through its material remains and impact on the landscape. Research in this subfield has at last "begun to embrace issues of cultural patterning and behaviours associated with the broader anthropology of conflict."[96]

The archaeology of the battlefields of the Western Front in World War One has been a particular concern of archaeologists Nicholas Saunders and John Schofield; both have been influential in establishing the archaeology of World War One as a subfield of landscape archaeology. Their work emphasizes the importance of new types of industrialized conflict, particularly trench warfare, and how it shaped and reshaped the topography of the battlefield along with the lives of those who were touched by war. For some archaeologists there is no distinct difference between conflict archaeology and battlefield archaeology; others, like Saunders, a leader in developing the standards and guidelines of modern conflict archaeology, believe this approach to archaeology should not be conflated with the earlier goals and methodologies of battlefield archaeology.[97] Battlefield archaeology, he claims, has its roots in military history and makes use of archaeology and material remains primarily to complement or confirm the textual records of military history research.[98] For Saunders the move towards conflict archaeology represents a break from the more traditional battlefield archaeology. Landscape archaeologist and anthropologist Sarah De Nardi observes that there has been "slow burning revolution in academic engagement with warfare" which has resulted in moving "histories of conflict from the unfashionable suburb of 'military history' to uptown locales like 'war studies' and 'First World War Studies'."[99]

Archaeologists who worked on the Gallipoli peninsula, such as Antonio Sagona, positioned conflict archaeology as a sub-discipline of archaeology but allowed it to include battlefield archaeology because, he claimed, in both fields "the original functions of weaponry and relics collected from the arena of war have been superseded. . . . These objects float where oral history, military history and archaeology intersect, where family history and global history cross."[100] Ultimately, conflict archaeology represents a more interdisciplinary approach to archaeology which "focuses on conflict as a multifaceted phenomenon, whose variety of physical traces possess multiple meanings that change over time. It is not restricted to battlefields, or to large-scale wars between nations, but embraces every kind of conflict and its diversity of social and cultural legacies."[101]

Moving the study of conflict archaeology beyond the terrestrial locations of the actual landscape, the newer practices and methodologies of maritime archaeology and the use of 3D digital topography and airborne laser scanning (ALS) to conduct "archaeology from above" are bound to shed new and interesting light on the Gallipoli peninsula and its archaeological heritage in future years.[102] The growing interest in more recent periods of archaeology and the impact of conflict on the landscape, studied through new technologies which combine historical research and its archival documentation with geospatial techniques, is contributing to what some scholars have referred to as a "spatial turn" in the humanities. While the archaeological excavations of historic battlefields, and even contemporary zones of conflict, are not new per se, the articulation of a particular intent to investigate geographies which have been impacted by violence and war, and the methodological practices used in these subdisciplines of archaeology have developed in significant ways in the past two decades and are beginning to shape the current survey and archaeological practices and methodologies used on the Gallipoli peninsula and in its coastal waters.[103]

Antiquities and artifacts of World War One

In addition to employing new technologies and new methodologies to understand the impact of World War One on the land and seascapes of the Gallipoli peninsula, archaeologists working in the region must continue to explore the many layers of the peninsula and diversify the interpretative strategies currently used to explain its pasts and present realities. Under the new administrative body of ÇATAB, and with the passing of the World War One centenary, the material culture from World War One is now considered to be of historic value and requires government protection and official museum registration. Increasing instances of vandalism and theft on the peninsula since 2014 have further contributed to a growing awareness among the different stakeholders of the heritage value of World War One material culture.[104]

Ultimately, linking the peninsula to its earlier, Ottoman, and more recent pasts—without neglecting the World War One narratives—has the potential to enrich the experiences of the archaeologist, the museum visitor, and the local residents of the peninsula.[105] When archeologists working on the peninsula "thicken the description" of the landscape through future surveys and excavations, they can also enrich the interpretative strategies currently used in the museums of the peninsula to explain the many pasts of that landscape, and to communicate the present realities of this region of Turkey. The more excavation that happens on the peninsula, the more layers of this complex past will emerge. The objects that are found in these archaeological excavations will need to be contextualized into richer and more diversified exhibition narratives, so it is crucial to begin considering how to develop these in existing and future museums in order to better elucidate the peninsula's diverse history and layers of landscape.

As the many archaeological pasts of the peninsula are explored and excavated in coming years, and museums are built to house newly discovered artifacts, those that were removed from the peninsula during and shortly after World War One could be returned to their rightful places. Objects removed by archaeologists during World War One from Elaious and during the excavations of the Protesilaus "tomb" (several which are in the collections of the Louvre) were not gifted or sold, as many antiquities had been to various European governments by the Ottoman sultan in return for political and military support.[106] As noted earlier, Ottoman laws protecting antiquities, restricting their export, and governing archaeological excavations undertaken by foreigners had been passed as early as 1869.[107] Halil Ethem Bey, director of antiquities from 1910 to 1931 for the Ottoman Empire during the war, and then in the department established by Mustafa Kemal Atatürk, had protested to Charles Picard, the director of the École Francaise d'Athènes, about the excavations they were conducting at Karaağaçtepe and insisted that the objects discovered there should be sent to the Istanbul Archaeology Museum.[108] The original catalog of these objects has since disappeared and it is therefore difficult to determine what percentage of these excavated objects was given to the Istanbul Archaeology Museum; but many artifacts excavated during the war by the French archaeologists working on the Gallipoli peninsula were taken to France. Efforts to repatriate artifacts taken during World War One have been less successful than for those taken during World War Two, as international legislation concerning the removal of cultural property during that second war is more developed and enforceable.[109] The Elaious artifacts were removed during a time of conflict and upheaval but also at a time when there was Ottoman legislation that prohibited this.[110] With plans for the construction of new museums on the peninsula that could safely house the objects from Elaious and Karaağaçtepe, there is a growing popular sentiment that it may now be time for some of these objects to return to Gallipoli.[111]

The value of ordnance and other memorabilia from the battles fought on the peninsula in World War One was recognized during the war, first as souvenirs by the soldiers who wished to keep a memory of the lands where they had fought and where many of their comrades had died. The material culture of the war was then recognized by Australia's historian C.E. Bean, who had spoken as early as 1915 about the need for a "museum in Australia in which the war relics would be preserved."[112] Other nations, including Canada and England, were beginning to establish their own war museums at that time, and by 1917 Bean had decided that, in addition to textual sources such as diaries, photographs, and sketches, the National Collection of the Australian War Memorial should include "the records in material-trophies, the things our men and units have actually used in battle or the German has used against them, with the mark of battle upon them."[113] By 1918 the Australian government had approved of the plans for a national museum dedicated to the Australian experiences on the battlefront and which would serve as the repository for many carefully conserved and curated objects, belonging to both the Ottoman and Allied troops, that were removed from the Gallipoli battlefields.[114] Turkey was very late to realize the heritage value of World War One ordnance and memorabilia and to define these as artifacts worthy of preservation and display in its museums.

As of 1981, the year that Turkey ratified the UNESCO Convention on the Means of Prohibiting and Preventing the Illicit Import, Export and Transfer of Ownership of Cultural Property of 1970, any artifact or ordnance dating to World War One or earlier which is a surface find or excavated on the Gallipoli peninsula, or in the waters surrounding it, or anywhere in Turkey for that matter, is legally protected.[115] Responsibility for enforcing this convention and prohibiting the removal of more cultural property from the Gallipoli landscape now falls to the new ÇATAB directorate.[116] Artifacts that have a cultural significance for a nation, as many objects dating to World War One do, need to be protected with legislation, and private collections of these types of objects now must be registered with local museums and their inventories monitored in the same way that private archaeological collections in the country are registered and controlled by the state.[117] As the 1915 centenary has passed, and these wartime objects are now at least a century old, there is an additional reason to protect these cultural assets and allow them to tell the rich and varied stories of the lands and the waters from which they have emerged.[118]

Notes

1 Joseph Vassal, *Uncensored Letters*, 127.

2 For a history of that project see Lucienne Thys-Şenocak, Carolyn Aslan, and Rahmi Nurhan Çelik, "Gelibolu Yarımadası Tarihsel Savaş Alanı Arkeolojisi Araştırmaları: Seddülbahir Kalesi Projesi (Research of the Historical and Battlefield Archaeology of the Gallipoli Peninsula: The Ottoman Fortress at Seddülbahir)," *TUBA-AR* 11

(Turkish Academy of Sciences Journal of Archaeology (Ankara, 2008), 107–20; Lucienne Thys-Şenocak and Carolyn Aslan, "Narratives of Destruction and Construction: The Complex Cultural Heritage of the Gallipoli Peninsula," in *Archaeology of Destruction*, ed. Lila Rokoczy (Cambridge: Cambridge University Press, 2008), 29–47.

3 These last two excavations began in 2014. For the 2017 report on ÇATAB activities see http://catab.kulturturizm.gov.tr/Eklenti/53920,2017-yili-6-aylik-faaliyet-raporu-catabpdf.pdf; the Çanakkale Museum also conducted excavations in 2008 at the Rumeli Mecidiye Barracks in the village of Kilitbahir. For the results of that excavation see Ö. Çavga and E. Çağman Esirgemez, "Çanakkale ili, Eceabat İlçesi, Kilitbahir Köyü, Rumeli Mecidiye Tabyası 2008 Yılı Kazısı". *18. Müze Çalışmaları ve Kurtarma Kazıları Sempozyumu*, 2009 no. 18 (Ankara 2010), 41–64.

4 Thys-Şenocak, "Divided Spaces, Contested Pasts", in *Public Nature*, 128.

5 Heinrich Schliemann, *Troja: Results of the Latest Researches and Discoveries on the Site of Homer's Troy and in the heroic tumuli and other sites, made in the year 1882: and a narrative of a journey in the Troad in 1881*. London: John Murray, 1884), 254–57.

6 Ibid., 260.

7 For a discussion of archaeology and espionage in World War One Ottoman lands, see David W.J. Gill, "*Excavating Under Gunfire: Archaeologists in the Aegean During the First World War*," *Public Archaeology*, 10, no. 4 (November 2011), 187–99.

8 Schliemann, *Troja*, 258.

9 Charles Picard and Adolphe J. Reinach, "Voyage dans la Chersonése et aux îles de la mer de Thrace," in *Bulletin de Correspondance Hellénique*, *36* (1912), 275–352; see also Catherine Valenti, "L'école française d'Athènes pendant la Grande guerre: une institution universitaire au service de l'Entente," *Guerres mondiales et conflits contemporains*, 204, no. 4 (2001), 5–14.

10 Gill, 189.

11 Angelika Waiblinger, "La ville grecque d'Éléonte en Chersonèse de Thrace et sa nécropole," *Comptes rendus des séances de l'Académie des Inscriptions et Belles-Lettres* 122.4 (1978), 843–57. Salaville Sévérien, "Ambassade de France a Constantinople. Fouilles du Corps d'occupation français de Constantinople exécutées de 1920 à 1923 par ordre du général Charpy, avec le concours technique de l'École française d'Athènes sous la direction de M. Ch. Picard, publiées par R. Demangel; Th. Macridy-Bey; E. Mamboury; Mme. M. Massoul; préface de Ch. Diehl; Fascicule I: *Le Tumulus dit de Protésilas*, par R. Demangel." *Échos d'Orient*, 26, no. 148 (1927), 494–7.

12 For a summary of the involvement of French archaeologists in Thrace and particularly the excavations in the Cape Helles region of the Gallipoli peninsula during World War One and from 1921–23, see Jean-Yves Marc, "*L'archéologie française en Thrace: A la mémoire des tirailleurs sénégalaise du 54émé regiment colonial*," *Actes du deuxieme symposium international des Etudes Thraciennes, Thrace ancienne* (1997), 473–506; Waiblinger, 843–57.

13 C.J. Mackie, et al., "Boundary and Divide: The Antiquity of the Dardanelles," *Anzac Battlefield*, 16–17; C.J. Mackie, "Archaeology at Gallipoli in 1915," *Philathenaios, Studies in Honour of Michael J. Osborne*, ed. Anastasios Tamis, C. Mackie, and S. Byrne (Athens: Anatypon, 2010), 213–25.

14 C.J. Mackie, et al., "Boundary and Divide", 15–16; C.A. Hutton, "Two Sepulchral Inscriptions from Suvla Bay," *The Annual of the British School at Athens* (1914–16) 21, 166–8.

15 The Australians who were proclaimed as missing in the Cape Helles region of the peninsula are commemorated on the Cape Helles Memorial near Seddülbahir. For information on the Roman remains at Lone Pine see R.A. Crouch, "Gallipoli Revisited," *Royal Historical Society of Victoria, Manuscripts Library*, MS 000123 Box 47/6 Unpublished 1924 as cited in Mackie, "Archaeology in Gallipoli in 1915," 20–21; Mackie et al., "Boundary and Divide," 7–8.

16 Vassal, 127–9.

17 Sergeant Edouard Dhorme was a member of the École Biblique et Archéologique in Jerusalem; the other three archaeologists graduated from École Française d'Athènes; Gill, 189–91.

18 Marc, "L'archéologie francaise en Thrace," *Actes du deuxième symposium international des Etudes Thraciennes. Thrace ancienne,* (1997), 473–506.

19 Fernand Courby, Joseph Chamonard, and Edouard Dhorme, "Corps expéditionnaire d'Orient. Fouilles archéologiques sur l'emplacement de la nécropole d'Éléonte de Thrace," *Bulletin de correspondance hellénique* 39 no.1 (1915), 135–240. doi:10.3406/bch.1915.3109; See also Matthieu René-Hubert, "Des hellénistes en guerre", *Revue historique des armées*, 261, (2010), 88–105. http://rha.revues.org/7111.

20 Ibid.; see also Mackie, "Archaeology in Gallipoli in 1915," 20–1; Mackie et al., "Boundary and Divide," 7–8.

21 Vassal, 142.

22 Ibid., 177.

23 Edmond Pottier, "Coupe attique d'Eléonte au musée du Louvre," *Bulletin de correspondance hellénique* 55 *(1)* (1931), 430–7.

24 Edhem Eldem, "From Blissful Indifference to Anguished Concern: Ottoman Perceptions of Antiquities, 1799–1869," *Scramble for the Past: A Story of Archaeology in the Ottoman Empire, 1753–1914*, ed. Zainab Bahrani, Zeynep Çelik, and Edhem Eldem (İstanbul: SALT, 2011), 281–2.

25 Ibid.

26 According to Sibel Özel, "Under the Turkish Blanket Legislation: The Recovery of Cultural Property Removed from Turkey," *International Journal of Legal Information*, 38, no. 2, Article 10 (2010), 179. http://scholarship.law.cornell.edu/ijli/vol38/iss2/10; "After the formation of the Republic of Turkey in 1923, the Turkish Civil Code was adopted in 1926. Even so, the 1906 Decree was maintained in full force and effect until 1973. The 1973 law preserved the same principle of state property. It again declared that all antiquities discovered in or on private or public lands in Turkey were the property of the state."

27 Morag Kersel, "The Changing Legal Landscape for Middle Eastern Archaeology in the Colonial Era, 1800–1930," *Pioneers to the Past 1919–20: American Archaeology in the Middle East*, ed. G. Emberling and Oriental Institute Museum (Chicago: Oriental Institute, University of Chicago, 2010), 85–6.

28 Robert Demangel, *Le Tumulus Dit de Protésilas* (Paris: de Broccard, 1926), 10–13.

29 During the campaign, the French dead had been buried in four main cemeteries and ossuaries around the peninsula: Le Cimetière Galinier, Le Cimetière de L'ambulance, Le Cimetière de la 1re Division, and Le Cimetière de la 2e Division; Cimetière Galinier was the cemetery for the headquarters of the French troops and was located

in Seddülbahir. See also www.gallipoli.com.tr/if_stones_could_speak/french_monument_cemetery.htm. The excavation at the Seddülbahir fortress in 2017 by ÇATAB revealed that several bones of the French soldiers had been overlooked by the French during the exhumation process. These human remains have since been excavated by ÇATAB and the French government notified so that they could undertake the reburial of these remains at the French cemetery at Morto Bay.

30 Mehmet Özdoğan, "Prehistoric Sites in the Gelibolu Peninsula," *Anadolu Araştırmaları,* 10 (1986), 54.

31 Demangel, 10–13.

32 Kılıç, 49–60.

33 For studies of this interaction see John Boardman, *The Greeks Overseas* (London: Penguin, 1964), 237; Walter Leaf, "The Military Geography of the Troad," *The Geographical Journal* 47 no. 6 (1916), 401–16; Benjamin H. Isaac, "The Greek Settlement in Thrace until the Macedonian Conquest," in *Studies of the Dutch Archaeological and Historical Society*, vol. 10, ed. Benjamin H. Isaac and J.G.P. Best (Leiden: Brill, 1986), 180.

34 Isaac, 177.

35 Isaac, 180; see also Xenophon, *Hellenica 2.1.22–32.*

36 John Bagnell Bury, Stanley Arthur Cook, and Frank E. Adcock, *The Cambridge Ancient History*, 2nd ed. (Cambridge: Cambridge University Press, 1933).

37 Demangel, 11–12.

38 According to Ertuğrul Ceylan, a lawyer who participated in the planning meetings for the monument's construction during the 1950s, no mention was made of the earlier archaeological site at Elaious. Interview with author on 10 April 2008, archived in Koç University Oral History and Memory Archive, Suna Kiraç Koç University Library (Thys-Şenocak collection).

39 This was an observation made by the author during the time of the expansion of that cemetery.

40 Thomas W. Laqueur, "Memory and Naming in the Great War," in *Commemorations: The Politics of National Identity*, ed. John R. Gillis (Princeton: Princeton University Press, 1994), 161.

41 For policies regarding the evolution of the commemorative procedures and regulations of the CWGC see Philip Longworth, *The Unending Vigil: A History of the Commonwealth War Graves Commission, 1917–1984* (S. Yorkshire: Leo Cooper, 2003).

42 Pemberton, p. 40.

43 Ibid., 40–1.

44 For a study of C.E. Bean's return to the Gallipoli battlefields see Janda Gooding, *Gallipoli Revisited: In the Footsteps of Charles Bean and the Australian Historical Mission* (Melbourne and London: Hardie Grant Publishing, 2009).

45 David William Lloyd, *Battlefield Tourism: Pilgrimage and the Commemoration of the Great War in Britain, Australia and Canada, 1919–1939* (Oxford: A&C Black, 1998), 184.

46 AWM38 3DRL 8042/51.

47 Ibid. Letter from C.E. Bean to Major J.L. Terloar 11 March 1919 written en route to Constantinople.

48 The surface survey of the Gallipoli peninsula conducted by David French, 27–28 August 1963, is not counted among the surveys listed here due to its brief nature and limited findings. See D. French, "Recent Archaeological Research in Turkey, Surface

Finds," *Anatolian Studies* 14, no. 37 (1964). The other surveys, here listed by director name and date, are as follows: Özdoğan, 1979–82; Bademli, 1997; Thys-Şenocak/Çelik, 1997–2002; Cameron et al. 2003.

49 L. Thys-Şenocak, "The Architectural Survey of the Ottoman Fortress of Seddülbahir," *17 Araştırma Sonuçları Toplantısı*, vol. 1 (Ankara: T.C. Kültür Bakanlığı Anıtlar ve Müzeler Genel Müdürluğu, 2000), 155–62.

50 For a detailed examination of the politics, past and present, of survey and archaeological work in Turkey see Nimet Pınar Özgüner, "Archaeological Entanglements: People, Places and Politics of Archaeology in Turkey", unpublished Ph.D. dissertation, (Boston University, 2015).

51 For these different categories of surveys and excavations see www.kulturvarliklari.gov.tr/TR,44150/kazi-ve-yuzey-arastirmalari-faaliyetleri.html. For the publications of the museum-led salvage excavations www.kulturvarliklari.gov.tr/TR,44763/muze-calismalari-ve-kurtarma-kazilari-sempozyumu-yayinl-.html. For the Trans-national Anatolian Natural Gas Pipeline see www.tanap.com/tanap-project/why-tanap. As mentioned in the text, the first officially approved archaeological excavation on the Gallipoli peninsula since World War One was the Seddülbahir fortress excavation conducted in the summer seasons of 2005 and 2006. This excavation has been categorized in the Ministry of Culture and Tourism archives as a salvage excavation because it was intended to provide information for the restoration and reusage of the site.

52 Mehmet Özdoğan, "A Surface Survey for Prehistoric and Early Historic Sites in Northwestern Turkey," *National Geographic Society Research Reports* 20 (1985), 517–41; also Özdoğan, "Prehistoric Sites in the Gelibolu Peninsula," 51–67.

53 Özdoğan, "A Surface Survey," 528–9.

54 Ibid., 530.

55 Among the many publications by Sayar on these epigraphic surveys are Mustafa Sayar, "Doğu Trakya," TÜBA-AR 10, (2007), 113–15; "Doğu Trakya Yüzey Araştırmaları 2005," XXIV. *Araştırma Sonuçları Toplantısı*, 29 Mayıs–2 Haziran 2006 Çanakkale (Ankara: 2007). vol. 2, 197–200; Doğu Trakya Yüzey Araştırmaları 2006, XXV. Araştırma Sonuçları Toplantısı, 28 Mayıs–1 Haziran 2007 Kocaeli (Ankara: 2008) vol. 2, 269–76. In addition to Sayar's work it is also important to note the cultural resources inventory compiled by M. Tombul, *Çanakkale Kültür Envanteri* (İstanbul: 2015), and the survey of Roman roads and milestones and the report on 'Republican Milestones' undertaken by David French of the British Institute of Archaeology in Ankara as his work included a survey of Gallipoli. http://biaa.ac.uk/ckeditor/filemanager/userfiles/rrmam%20vol.%203%20milestones.pdf.

56 R. Bademli, "Gelibolu Yarımadası Barış Parkı Projesi Uluslararası Fikir ve Tasarım Yarışması ve Sonrası," *Mimarlık*, (36) 283 (1998), 25–6.

57 Ibid.

58 http://vitruvius.arch.metu.edu.tr/gallipoli/gallipoli_turkish.html. For the report see: R. Bademli, ed. (2001) *Gelibolu Yarımadası Tarihi Milli Parkı (Barış Parkı) Uzun Devreli Gelişme Planı (UDGP) Çalışmaları*, vol. 2: Esaslar ve Tutumlar (Basılmamış Rapor), ODTÜ, Gelibolu Yarımadası Barış Parkı Planlama ve Danışma Bürosu, AGUDÖS Proje No 99.02.02.03, Ankara.

59 David Cameron and Denise Donlon, "A Preliminary Archaeological Survey of the ANZAC Gallipoli Battlefields of 1915," *Australasian Historical Archaeology 23* (2005), 131–8.

60 Antonio Sagona, "An Archaeology of the ANZAC Battlefield," *Humanities Australia 6* (2015), 34–46.

61 Antonio Sagona and Jessie Birkett-Rees, "Battlefield Archaeology: Gallipoli," *Anzac Battlefield*, 96.

62 Simon Harrington, "Human Remains," *Anzac Battlefield*, 93–4.

63 Ibid., 94.

64 Gooding, 112.

65 Onur Özbek, "Hamaylıtarla Reconsidered: A Neolithic Site and Its Environmental Setting in Southern Turkey," *Anatolia Antiqua* 18 *(1)* (2010), 1–21.

66 Ayşe Çaylak-Türker, "The Byzantine Architectural Sculptures with Figure at Hellespont," *Hacettepe Üniversitesi Edebiyat Dergisi,* vol. 23 no. 1 (2006), 173–90.

67 Göksel Sazcı, "Maydos Kilisetepe Settlement Mound," in *City of Legends and Epics Çanakkale*, ed. F. Özden (İstanbul: Yapı Kredi Press, 2012), 227–30.

68 Ibid., 227–8.

69 Ibid., 229; see also Göksel Sazcı and Carolyn Aslan, "Across the Hellespont: Maydos (Ancient Madytos), Troy and the North-eastern Aegean in the Late Eighth–Early Sixth Century BC," *Annual of the British School of Athens* 111 (2016), 121–62.

70 Lucienne Thys-Şenocak, Ottoman Women Builders: The Architectural Patronage of Hadice Turhan Sultan (Aldershot: Ashgate, 2006), 113–15.

71 For a discussion of these restoration decisions see Lucienne Thys-Şenocak and Carolyn Aslan, "Narratives of Destruction and Construction: The Complex Cultural Heritage of the Gallipoli Peninsula," in *Archaeology of Destruction*, ed. Lila Rokoczy (Cambridge: Cambridge University Press, 2008), 29–47. See also, A. Özsavaşcı, Lucienne Thys-Şenocak, G. Tanyeli, R.N. Çelik and Çanakkale Savaşları Gelibolu Tarihi Alan Başkanlığı, "Rölöve, Restitüsyon, Konservasyon Projelerinden Restorasyon Uygulaması ve Yeni Kullanım Süresinde Seddülbahir Kalesi," *Kargir Yapılarda Koruma ve Onarım Semineri VIII Conference Proceedings*, KÜDEB, İBB (İstanbul, 2016) 236–61.

72 The lives of some of these Turks who had migrated to the peninsula from Romania and Bulgaria in the 1930s were explored and documented during the oral history project we conducted at Seddülbahir, Alçıtepe and Kumkale from 2000 to 2002. See L. Thys-Şenocak and I.C. Cenker, "Moving Beyond the Walls: The Oral History of the Ottoman Fortress Villages of Seddülbahir and Kumkale," *Oral History and Public Memories*, ed. P. Hamilton and L. Shopes (Philadelphia, Temple University Press, 2008), 65–86.

73 Uzi Baram and Lynda Carroll, ed., *A Historical Archaeology of the Ottoman Empire: Breaking New Ground* (New York: Springer Science & Business Media, 2006).

74 The regions where the 2015 Körpe survey took place include Kilye Ovası, Poyraz Tepe, Güney Yamaçları, Bigalı Ovası Ece Limanı, Büyük Kemikli Burnu, Kireçtepe Batısı, B. Anafartalar Ovası, Softa Tepe, K. Anafartalar Ovası, Yılgın Tepe, Anafartalar Ovası, Lalebaba, Büyük Anafartalar Köyü, Çamlıtekke Mevkii, Seddülbahir. www.ttk.gov.tr/wp-content/uploads/2016/11/9-Gelibolu.pdf; and R. Körpe, "Çanakkale Savaş Alanlarında Yapılan Arkeolojik Araştırmalar." *100. Yılında Çanakkale Savaşları Uluslararası Kongresi/ International Congress on Battles of Gallipoli on the Centenary*, ed. A. Akengin and O. Neçare, (Ankara: Atatürk Araştırma Merkezi Yayınları, 2018), 773–94.

75 See updates of the Archaeological Settlements of Turkey Database (TAY) at www.tayproject.org/trhome.html and the publications of the conference proceedings

of the research, survey and excavations held in Turkey each year (Kazı Sonuçları Toplantıları) by the Turkish Ministry of Culture and Tourism. www.kulturvarliklari.gov.tr/TR,44760/kazi-sonuclari-toplantilari.html.

76 For a history of that project see Lucienne Thys-Şenocak, Carolyn Aslan, and Rahmi Nurhan Çelik, "Gelibolu Yarımadası Tarihsel Savaş Alanı Arkeolojisi Araştırmaları: Seddülbahir Kalesi Projesi (Research of the Historical and Battlefield Archaeology of the Gallipoli Peninsula: The Ottoman Fortress at Seddülbahir)," *TUBA-AR* 11 (*Turkish Academy of Sciences Journal of Archaeology*) (2008), 107–20.

77 Antonio Sagona, "An Archaeology of the ANZAC Battlefield," *Humanities Australia 6* (2015), 34–46.

78 M. Cleary, "Anzac Gallipoli Archaeological Database," *Anzac Battlefield*, 246–50.

79 Antonio Sagona, "An Archaeology of the ANZAC Battlefield," *Humanities Australia 6* (2015), 34–46.

80 Antonio Sagona and Jessie Birkett-Rees, "Battlefield Archaeology: Gallipoli," *Anzac Battlefield*, 95.

81 Aydın Büyüksaraç, Cahit Çağlar Yalçıner, Yunus Levent Ekinci, Alper Demirci, and Mehmet Ali Yücel, "Geophysical Investigations at Ağadere Cemetery, Gallipoli Peninsula, NW Turkey," *Australian Journal of Forensic Sciences* 46, no. 1 (2014), 111–23.

82 Ibid. See also www.haberler.com/agadere-sehitlik-alani-olarak-tescil-edildi-3085524-haberi/. For information on the first prize architectural competition for the Ağadere complex http://www.arkiv.com.tr/proje/1-odul-gelibolu-yarimadasi-tarihi-milli-parki-agadere-hastanesi-ve-sehitligi-fikir-projesi-yarismasi/741.

83 UNESCO *Convention on the Protection of the Underwater Heritage*, Paris, 2001; (http://unesdoc.unesco.org/images/0012/001260/126065e.pdf).

84 S. Kolay, O. Taktak, S. Karataş, and M. Atabay, "The Salvage and Sale of Wrecks at the Dardanelles," *Echoes from the Deep: Wrecks of the Dardanelles Campaign* (İstanbul: Vehbi Koç Foundation/Ayhan Şahenk Foundation, 2013), 264–9. See also Resmi Gazete, no. 3792, 25 December 1937, and BCA 030.10.00.00.62.415.13.2.

85 S. Kolay et al., 269.

86 The Turkish Ministry of Culture and Tourism permits for underwater research in the Çanakkale region were granted to Harun Özdaş in 2007.

87 For the proceedings of "Underwater Cultural Heritage United Nations from World War I," see www.unesco.org/culture/underwater/world-warI.pdf.

88 Selçuk Kolay and Savaş Karatay, "Echoes from the Deep: Wrecks of the Dardanelles Campaign," in *Underwater Cultural Heritage from World War I: Proceedings of the Scientific Conference on the Occasion of the Centenary of World War I* (Bruges, Belgium: UNESCO Publishing, 2015). For a list of wrecks of Ottoman and Allied forces which sank between 1914 and 1918, see page 110 from H. Barış Özalp, "Çanakkale Savaşlarına Sualtından Bir Bakış (A Look From Underwater to the Çanakkale Warfares)," Çanakkale Araştırmaları Türk Yıllığı/ The Turkish Yearbook of Gallipoli Studies, 8–9 (2010), 91–110.

89 One such project where opinions differed concerning excavation/salvage vs. only documentation was "AE2 The Silent ANZAC," an Australian British E-class submarine which managed to navigate through the Dardanelles and into the Sea of Marmara prior to 25 April 1915. See Tim Smith, "AE2 The Silent Anzac: Australia's Lost Gallipoli Submarine," Paper 34, *Iron, Steel and Steamship Archaeology, Proceedings of 2nd Australian Seminar,* (Melbourne and Sydney: Freemantle, 2006), 139.

90 S. Kolay, O. Taktak, S. Karataş, and M. Atabay, *Echoes from the Deep: Wrecks of the Dardanelles Campaign* (İstanbul: Vehbi Koç Foundation/Ayhan Şahenk Foundation, 2013), 27.

91 Ibid., 29.

92 For details about this expedition see "Report of the Expedition: Gallipoli 2011" http://general.gpe.pl/ww/wp-content/uploads/2016/08/EN_Raport_Gallipoli_2011.pdf.

93 For the 20 June 2010 media release of this project by the NSW government see, "Gallipoli Heritage Survey Uncovers War Secrets"; http://www.planning.nsw.gov.au/~/media/Files/DPE/Media-Releases/2010/June/20062010-gallipoli-heritage-survey-uncovers-war-secrets.ashx.

94 Tim Smith, "Project Beneath Gallipoli," *The Gallipolian*, 126, (2011), 37–45.

95 See Chapter 6 of this book for an analysis of this report.

96 Jessie Birkett-Rees, "Geospatial Science and the Archaeology of the First World War: Context for Conflict," *Geospatial Science Research* 2 (2012). n.p.; http://ceur-ws.org/Vol-1328/GSR2_BirkettRees.pdf.

97 For an introduction to the methodologies of First World War archaeology see Nicholas J. Saunders, *Killing Time: Archaeology and the First World War* (Stroud: Sutton Publishing, 2007).

98 For further discussions concerning battlefield vs. conflict archaeology see Douglas D. Scott, Richard A. Fox, Melissa A. Connor, and Dick Harmon, *Archaeological Perspectives on the Battle of the Little Bighorn* (Norman, OK: University of Oklahoma Press, 2013).

99 Sarah De Nardi, *The Poetics of Conflict Experience: Materiality and Embodiment in Second World War Italy* (Oxford and New York: Routledge 2017).

100 Sagona et al., "An Archaeology of the Anzac Battlefield, Humanities Australia", 35.

101 For this definition see https://modernconflictarchaeology.com/about/.

102 For an example of the use of ALS at Gallipoli see J. Birkett-Rees, "Aerial Perspectives on Archaeological Landscapes: The Anzac/Arıburnu Battlefields of Gallipoli, Turkey," in *Conflict Landscapes and Archaeology from Above*, ed. Birger Stichelbaut and David Cowley (Oxford and New York: Routledge, 2016); see also the Archaeo-landscapes Europe Project, www.arcland.edu.

103 *Matters of Conflict, Material Culture, Memory and the First World War*, ed. N. Saunders (London: Routledge, 2004); *Contested Objects: Material Memories of the Great War*, ed. N. Saunders and P. Cornish (London: Routledge, 2008); *Beyond the Dead Horizon: Studies in Modern Conflict Archaeology*, ed. N. Saunders (Oxford: Oxbow, 2012). A useful introduction to Battlefield Archaeology is *Fields of Conflict: Battlefield Archaeology from the Roman Empire to the Korean War*, ed. D. Scott et al. (Westport CT: Praeger Security International, 2007).

104 Presentation of Çanakkale Wars and Gallipoli Historical Regional Directorate 23 Feb. 2017, Çanakkale. Some of the objects that were stolen were the copper letters "ANZAC" which were mounted on the commemorative wall of the ANZAC Bay Ceremonial Area on 28 October 2014 (www.geliboluyuanlamak.com/544_gelibolu-yarimadasinda-utanc-verici-olay-the-shameful-event-in-gallipoli-peninsula.html). On 2 March 2017 it was discovered that during the 2015 restoration of the Kilitbahir Namazgah and the Mecidiye Batteries, close to 9,000 World War One bullets had been stolen. See www.milliyet.com.tr/canakkale-savasi-ndan-kalan-gundem-2406132/Milliyet.com.tr, "Çanakkale Savaşı'ndan kalan binlerce mermi kovanı çalındı" (2 March 2017).

105 The investigation into the different eras of the Ottoman past is also facilitated by developments in Ottoman archaeology which has developed substantially in the past two decades. For an early collection of articles specifically dedicated to the study of Ottoman archaeology see Uzi Baram and Lynda Carroll, ed., *A Historical Archaeology of the Ottoman Empire: Breaking New Ground* (New York: Springer Science & Business Media, 2006); Ibolya Gerelyes and Gyöngyi Kovacs, *Archaeology of the Ottoman Period in Hungary* (Budapest: Hungarian National Museum, 2003). For Ottoman archaeology in the territories of the Black Sea region see C. Finkel and V. Ostapchuk, "Outpost of Empire: An Appraisal of Ottoman Building Registers as Sources for the Archaeology and Construction History of the Black Sea Fortress of Özi," *Muqarnas* 22 (2005), 150–88.

106 This point has been noted by James Goode, *Negotiating for the Past: Archaeology, Nationalism, and Diplomacy in the Middle East, 1919–1941* (Austin: University of Texas Press, 2007). See also Ayşe Özdemir, "A History of Turkish Archaeology From the Nineteenth Century to the End of the One-Party Period". Unpublished MA thesis, (Boğaziçi University, 2001).

107 James Goode, "Archaeology and Diplomacy in the Republic of Turkey, 1919–39," in *Turkish-American Relations: Past, Present and Future*, ed. Mustafa Aydın and Çağrı Erhan (London: Routledge, 2004), 49–65.

108 Goode, *Negotiating for the Past*, 25–6.

109 For an extensive discussion of this topic see A. Vrdoljak, *International Law, Museums and the Return of Cultural Objects* (Cambridge: Cambridge University Press, 2006).

110 See www.historynet.com/the-war-over-plunder-who-owns-art-stolen-in-war.htm.

111 See for example Mustafa Yurdakul's blog: https://turkiyekayipkulturhazineleriniariyor.wordpress.com/yazilar/canakkale-muharebeleri-esnasinda-gelibolu-yarimadasindan-calinan-arkeolojik-degerler/.

112 Gooding, *Gallipoli Revisited*, 14.

113 Ibid.

114 Peter Pedersen, *Anzac Treasures: The Gallipoli Collection of the Australian War Memorial* (Sydney: Murdoch Books, 2014).

115 http://portal.unesco.org/en/ev.php-URL_ID=13039&URL_DO=DO_TOPIC&URL_SECTION=201.html. The instruments agreed to by Turkey are: UNESCO Convention 1970 in force since 21 July 1981; UNESCO Convention 1954 and Protocol in force since 15 March 1966; European Convention on Offences relating to Cultural Property 1985, signed on 26 September 1985.

116 This was enacted on 28 June 2014 (29044 Resmi Gazete) See http://teftis.kulturturizm.gov.tr/TR,102525/http://teftis.kulturturizm.gov.tr/TR,102525/canakkale-savaslari-gelibolu-tarihi-alan-baskanligi-kur-.html-.html. Information about 2000 of the artifacts from the JHAS Gallipoli battlefield survey can be accessed in the AGAD (Anzac Gallipoli Archaeological Database) database accessible at http://agad.anzac.unimelb.edu.au/.

117 Ibid. All private collections of archaeological objects in Turkey must adhere to the legislation as outlined in the Ministry of Culture and Tourism's Law on the Conservation of Cultural and Natural Property (2863). For an English translation of this law see www.kulturvarliklari.gov.tr/TR,43249/law-on-the-conservation-of-cultural-and-natural-propert-.html.

118 L. Thys-Şenocak and Elifgül Doğan, "Archaeology, Museums and Tourism on the Gallipoli Peninsula: Issues of Human Remains, Ordnance, and Local Decision-

Making/Gelibolu Yarımadası'nda Arkeoloji, Müzeler ve Turizm: İnsan Kalıntıları, Mühimmat ve Yerel Karar Alma," *Heritage in Context II* (İstanbul: German Archaeological Institute (DAI), 2018) 289–313.

4 Commemoration begins for the Commonwealth and its Allies

In November 1918, while the diplomatic circles of Europe haggled over who owned what on the Gallipoli peninsula, the British Army Graves Registration Unit (GRU) had begun the grueling task of identifying and marking the places where bodies had fallen, and where they had been hastily buried. Leading the unit was Charles Bean, Australia's official historian of the Gallipoli campaign and one of the first officers from the Allied Forces to return to the Gallipoli peninsula soon after the war had ended. Bean, the Australian engineer Lieutenant Cyril Hughes, New Zealander Captain Clifton Bigg-Wither, Tasman Millington, and several members of the GRU were assigned to work first on the ANZAC section of the peninsula.

When the Treaty of Sèvres fell through in 1922, new arrangements were made with the government of Mustafa Kemal Pasha, İsmet Pasha, and their representatives about the nature and extent of claims to the land the Allies had requested for cemeteries. Concerned about the compromise that ceding the requested and strategic land to a former enemy implied, İsmet Pasha claimed that the Turks would treat the "remains of heroes with boundless respect, but . . . it was repugnant to them to mix this question up with the interests of the living."[1] At the conference in Lausanne on 24 July 1923 an agreement was finally reached concerning the extent of ownership, use, maintenance, and access to the Commonwealth burial sites on the peninsula.[2] By 1926, 31 cemeteries and five large memorials had been built by the Commonwealth War Graves Commission (CWGC). As stated in the Commission's brochure:

> After mammoth battlefield clearance operations 31 Commonwealth cemeteries were left containing 19,000 graves of which 6,000 were identified. A further 2,500 of the dead believed to be buried among the unidentified are commemorated in the cemeteries by Special Memorials bearing an inscription to this effect. The remainder of those buried in unknown graves, or whose remains were never found, make up the 27,000 named on six memorials to the missing on Gallipoli.[3]

The Imperial War Graves Commission (IWGC) had taken over the initial reconnaissance work of the GRU and would evolve into the Commonwealth War

Graves Commission (CWGC) which today maintains the Commonwealth cemeteries and memorials on the Gallipoli peninsula and throughout the world. In less than a decade after the evacuation of the peninsula by the Allied forces, a significant portion of the landscape of Gallipoli had been reorganized, renamed, and replanted. The CWGC's interest in the peninsula was largely concentrated in the areas where the Allied graves were located and attention was paid to completing infrastructure projects such as roads, small railways, nurseries, and wells, all which were needed to facilitate the building and landscape work for the many commemorative projects envisioned for the former battlefields. Between 1916 and 1926, how and where the Allied dead on the Gallipoli peninsula were to be buried and commemorated was decided, like the diplomacy of the war, far from the peninsula—in the British Foreign Office in London; in the Parliament and Commonwealth Offices of Australia; and in the architectural studios of the British and Australian architects who designed the monuments, tombstones, cemeteries, and gravesites for the Gallipoli battlefields.

The question of repatriation

Approximately 44,150 Allied troops had lost their lives on the peninsula; of these, many died at sea, or their bodies were never found, or were found in such a way as to be unidentifiable.[4] Compared to the Gallipoli battlefields, war on the Western Front had stretched out for longer, and the number of casualties was far higher. The British government, therefore, had decided within the first months of the war that the state should take over the responsibility for the burial and commemoration of its war dead. On the European battlefields where trench warfare had continued for months in the same spaces, the dead had literally fallen one on top of another, making the process of exhumation and identification (not to mention the problems of transporting individual bodies away from the front) particularly difficult. During the actual fighting, repatriating bodies for burial was a difficult logistical challenge. When the war finally ended, the sheer cost of bringing home what remained of bodies (assuming these were even able to be identified) led to a controversial, unpopular, and frequently contested decision by the British government against the repatriation of the Commonwealth war dead.

Establishing cemeteries outside of Britain, close to the battlefields where the soldiers had been killed, was deemed more practical for those soldiers who had lost their lives during battles fought on the European continent. The allocation by France to Britain of land for 200 cemeteries initiated what Laqueur has called "the era of the common soldier's name" and "a historically unprecedented planting of names on the landscapes of battle."[5] Thereafter, "the great ribbons of war cemeteries across Europe and the other theaters of war, with their neat individual grave markers, were thus genuinely new and were, in fact, regarded by their proponents as explicit rejections of the Victorian aesthetic of death."[6]

The concept of the state providing for the burial of all soldiers in a manner decided by the state, while based on egalitarian ideals promoted by the government, also separated the dead from their families, local communities, and churches.

Those who fell in battle were now laid to rest in collective graves with their comrades, far from the family plot in the home town where their ancestors had been buried. Personal grief and public honor were to be divided into two separate spaces. Grieving in private could be done at home, but the burials of the heroes who fell in war fighting for a common cause and for a common nation were to be standardized; not to be commemorated in what one British official, Lieut. Colonel Sir Frederic Kenyon, described as, "a jumbled mass of individual monuments of all sorts and sizes and of all variety of quality."[7]

After World War One, throughout Europe headstones were planned to have a uniform design, laid out on a continuous flat surface, and graves would not be articulated as separate mounds. Special landscaping signaled that the space was an organized cemetery, not a garden with meandering paths or romantic nooks reminiscent of the Victorian era landscape aesthetics. Burying the dead in rows and with uniform markers created an impression in death of the kind of discipline and order that was expected of the soldiers while they were alive. Everyone, regardless of rank, had the same size plaque or headstone; there would be no way to tell a Scot from an Irishman, a wealthy aristocrat from a poor carpenter. No hierarchy would intrude, as all had made the ultimate sacrifice for their homeland. Each of the soldiers who had served the state to his last breath remained the property of the British state forever.

The commemoration of the dead on or near the places where they had fallen on the Western Front made sense from a purely logistical perspective. It made even more sense, at least to the British Ministry of War, when the burials and bodies were at an even greater distance, as in the case of Gallipoli. After the evacuation of the Allied troops from the Gallipoli peninsula in December 1914–January 1915 it was clear within a few months that someone needed to identify the remains of the Allied dead, and if possible arrange for proper burials of the soldiers who had died on the peninsula. Britain's GRU, and its founder Sir Fabian Ware, took the initiative. Like the decisions made against repatriating British dead on the European mainland, leaving the remains of British dead on the Gallipoli peninsula was unpopular with many who wished the grave of a family member to be located closer to home, or who wished to establish a more personalized memorial for a family member.

Travel to Belgium or France to visit the grave of a husband or son was possible for many who could afford the three-day pilgrimage tours organized for this purpose by Thomas Cook and the other battlefield tourism agencies that emerged at the end of the war. But travel to the shores of Gallipoli was another matter. The distance and expense, and the ambivalent legal status of the graves on the peninsula (which would remain undetermined until the 1923 Treaty of Lausanne), made the journey to Gallipoli prohibitive for many relatives living in Great Britain, Australia, or New Zealand.

The question of repatriating bodies was decided differently by the French, the Allied country which, after the British Commonwealth, had seen the greatest number of dead in Gallipoli. On 28 September 1920 the French government decided that families of the war dead had the right to claim the bodies of their loved ones and have them brought home for burial.[8] It also agreed either to pay for the transport of the remains of soldiers who had died abroad or to subsidize a voyage for immediate family members who opted to bury their loved ones in the French cemeteries located near the battlefields where they had been killed.[9]

In spite of much popular sentiment against the repatriation ban, the British government held fast to its initial decision, and no exceptions were to be made for families who had lost a son, husband, or father on the Gallipoli peninsula. Commemoration was to be done collectively, as a nation. In the words of the Commission, "those who had given their lives are members of one family, and children of one mother who owes to all an equal tribute of gratitude and affection, and that, in death, all from General to Private, of whatever race or creed, should receive equal honor under a memorial which should be the common symbol of their comradeship and of the cause for which they died."[10]

Of course there were a few exceptions made, and one of these was at Gallipoli. While his remains were not repatriated, one Commonwealth soldier on the peninsula did have an isolated grave: Lieutenant-Colonel Charles Doughty-Wylie. He had been killed during the siege on 25–26 April 1915 to capture the fortress at Seddülbahir on Cape Helles and his many acts of bravery had earned him a Victoria Cross; but his individual burial plot on the peninsula was due to his wife, Lady Lilian Doughty-Wylie, who had managed to purchase a hundred square meters of land for a memorial gravesite for her husband, thereby allowing her to deviate from the rules of commemoration established by the Commission.[11]

Far more common was the experience of families such as the Burtons from St. Mary's Island, Cornwall, New Zealand. When Mrs. Burton asked the High Commissioner in a letter dated 9 August 1921 to use her check of fifty pounds to place "a suitable stone over my husband Frank Burton's grave," she received a negative reply.[12] Frank Burton, a sapper with the New Zealand Expeditionary Force who was buried in the Shrapnel Valley Cemetery on the Gallipoli peninsula, would be commemorated with a headstone but in the same manner as all the others who died with him and at the discretion of the government.[13] It was not suitable, replied the High Commission, or "practical to allot the gift you contemplate making to any specific grave or memorial."[14] The rank, name of the soldier, his regiment, and date of death would be standard for each headstone. If relatives wished to contribute something to the headstone, only a short epitaph or prayer, approved by the Commission, could be added.[15]

Once the standards and guidelines for memorializing the war dead were decided, headstones of uniform design and dimensions were created for Gallipoli, as they were throughout the cemeteries of Europe. Names of the dead had to fit within approved templates, and the epitaphs or prayers could not be too long for the headstones that marked the graves of the Allied dead in these distant lands. No excessive number of initials of those commemorated were to be placed after their names, as Mrs. Whittall of Birmingham discovered when she was asked on 25 November 1925 by the Director of Records, Committee of Imperial Defense to confirm, "Whether it is correct that the late Cpl. Whittal, of the Warwickshire Yeomanry, had six initials "W.J.G.P.A.L."; and if so, whether you desire that they should all be used in engraving his name on the Commission's Memorial at Cape Helles, Gallipoli?" In fact a decision had already been made in London concerning this extensive row of intitials; "the number of letters concerned is too many to be engraved."[16] Mrs. Whittall would have to do away with some of those if her husband's name was to appear on the Commission's Memorial at Cape Helles, Gallipoli.

Many British families, long after the names of the dead and prayers for them had been etched into tombstones, remained highly critical of the CWGC's restrictive efforts to standardize World War One burials. Public criticism of the Commission extended to the choice of words selected by the CWGC for the collective epitaphs inscribed on the peninsula's commemorative monuments. In a letter to the IWGC dated 23 February 1927, its director, Henry Chettle, received a complaint from a relative of a deceased soldier. The problem concerned the standard epitaph that had been chosen for some of the commemorative monuments on the Gallipoli peninsula: "Their glory shall not be blotted out." The phrase had been composed by Rudyard Kipling, one of Britain's most well-respected writers, who had lost a son on the World War One battlefields of Western Europe. However, Kipling's fame did not impress everyone. One father wrote that the words chosen by the lauded poet were "ugly" and suggested an alternative: "Their glory shall abide forever."[17] Kipling had been asked on several occasions to help with the wording of the Gallipoli epitaphs and was particularly sensitive to the terms used, as is evident in some of the correspondence between him and Chettle in 1926. Describing one group of Commonwealth soldiers who had died at sea in the Cape Helles sector, Kipling pointed out that including the word "hired" to indicate that a few of the boats in the campaign had been paid for their work was rather insensitive. The original text proposed by the Commission was, Kipling wrote:

> clear enough but rather severe in outline, and the inclusion of the fact that two or three of the transports were "hired" makes it a little bleaker than is needed. I suggest, if it does not interfere with any Departmental record, knocking out "hired".

In the same letter Kipling suggests that rather than write on the epigraph, "all of whom have no other grave than the sea," a passage from Job 12.22 would be more suitable: "He discovereth deep things out of darkness, and bringeth out to light the Shadow of Death." Kipling concluded his correspondence with an apology to his reader for his poor typing skills (Figure 4.1).[18]

As part of its efforts to alleviate grief and bridge the distance between the living and the dead, the Commission developed a policy of photographing each of the burial sites on the Gallipoli peninsula. While it was only after the Armistice was signed in 1918 that it was it possible for official commemorative and burial procedures such as photographing the graves to begin, unofficial pictures of the graves of Australians who had died in Gallipoli appeared as early as 1916 in *The Queenslander* newspaper. In the 8 January 1916 issue an image of a gravesite on

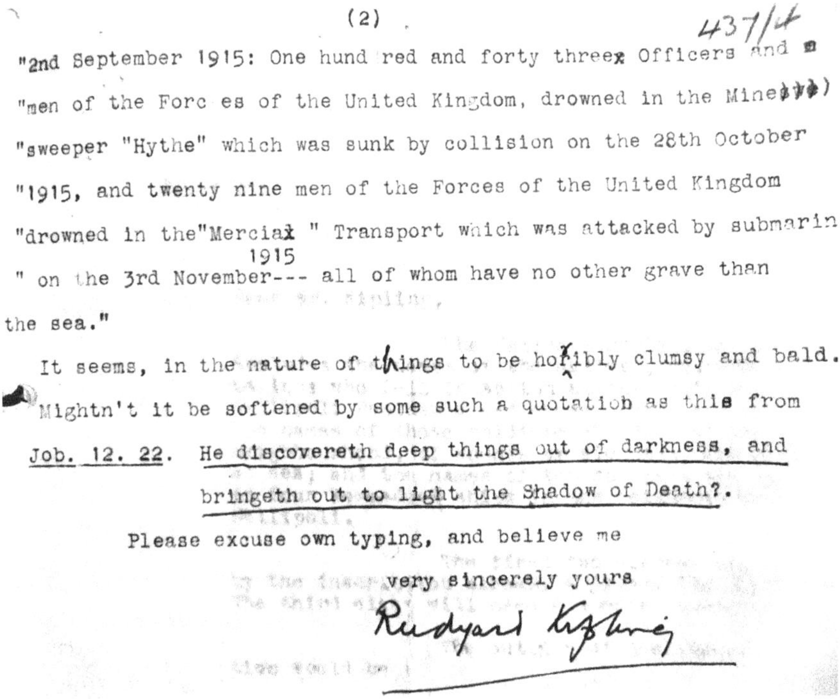

(2) 437/4

"2nd September 1915: One hundred and forty three Officers and
"men of the Forces of the United Kingdom, drowned in the Mine
"sweeper "Hythe" which was sunk by collision on the 28th October
"1915, and twenty nine men of the Forces of the United Kingdom
"drowned in the "Mercia " Transport which was attacked by submarin
" on the 3rd November 1915--- all of whom have no other grave than
the sea."

It seems, in the nature of things to be horribly clumsy and bald.
Mightn't it be softened by some such a quotation as this from
Job. 12. 22. He discovereth deep things out of darkness, and
bringeth out to light the Shadow of Death?.

Please excuse own typing, and believe me

very sincerely yours

Rudyard Kipling

Figure 4.1 Letter written by Rudyard Kipling concerning the choice of words for the epitaph on the Cape Helles monument, 24–6 March 1926.

Credit: Courtesy of the CWCG archives

the peninsula is captioned "Graves Where our Heroes are Buried."[19] Traveling to the burial sites was impossible for most families in Australia and New Zealand, so photographs became a tangible way a bereaved family could connect their imagination of a distant landscape to the place where a loved one lay.

Henry Muller's letter of 20 September 1920 to the Secretary of the IWGC was one of many similar requests by families seeking photographic evidence of a burial place on the peninsula. He wondered: "if it would be possible to get a Photograph, through your kind services, of the Grave of my Son Major G.F. Muller in Gallipoli."[20] Muller's father added to his polite request that he had heard this was something which was being done and he was ready to pay for that service. Muller had "noticed lately in the 'GLOBE and LAUREL,' the organ of the Marines, an extract from a letter from a Sergeant, I think, of one of the Visiting Detachments in Gallipoli, stating that the Graves of Col. Bendyshe and Major G.F. Muller had been located and left undisturbed. I think it is possible therefore, that there may be some Photographs of these Graves obtainable and if possible I should much like to have one. Of course I wish to pay for it. I am, Dear Sir, Yours very truly, Henry Muller."[21]

Because the decisions made in London against the repatriation of bodies extended to all the Commonwealth soldiers who had died in Gallipoli, by April 1917 plans had begun for an extensive network of cemeteries on several different points of the peninsula.[22] The governments of Australia and New Zealand, concerned about the care of their dead on the Gallipoli peninsula, were quick to take action and by November 1918 the GRU had established an office in the fifteenth century Ottoman fortress of Kilitbahir, just across the Dardanelles from present-day Çanakkale.[23] Their mission was to "proceed with the temporary marking of the graves on Gallipoli, the temporary enclosing of the cemeteries and generally with the checking of burial information available and searching for further graves."[24] Lieutenant Cyril Hughes was appointed as the Chief Officer of the GRU and worked with this group for two years identifying bodies and bringing the remains of soldiers to designated central burial areas.[25] Charles Bean began working with Hughes and several of members of his staff on the ANZAC battlefields early in 1919. As noted in the previous chapter, Bean was also intent upon collecting wartime artifacts or "trophies" to bring back to Australia. Salvaging ordnance and other material evidence of the war to take back to the museums in Australia, while insuring through commemorative monuments and cemeteries that evidence of Australian participation in that war remained on the peninsula, was of great importance to Australia's official historian of the war. In his correspondence written to Hughes at this time, the importance of the Gallipoli campaign for the Australians, and its role in shaping the nationalist narrative that he would be helping to construct for Australia, was already evident.

In an interesting request to Hughes, Bean asked: "Can you keep the Baby 700, Chunuk Bair, and Pine Ridge graves where they were; the story of the men who got to those places will figure prominently in the first two volumes of the history which will be published during this year."[26] This early commemorative campaign undertaken by the British, Australians, and New Zealanders had to deal with much uncertainty, including the fact that the peninsula itself was still officially "enemy territory" until its status was resolved in the 1923 Lausanne Treaty.

The work of the Commission was interrupted by difficult winters and continuing political instability, such as the 1922 Turkish War of Independence which uprooted and displaced the population, a portion of which had constituted the team's work force. The Gallipoli peninsula presented different obstacles than the flatter fields of Flanders and Verdun, and the process of identifying and exhuming bodies from here was an enormous challenge.[27] Noting how difficult it would be to exhume so many of his dead compatriots from the steep ridges and deep crevices of the Gallipoli landscape, Bean had recommended that "all cemeteries remain on their present site, fenced with salvaged material, the paths made up and the cemeteries planted with small Australian trees, not altering the appearance of the battleground . . ." He also proposed that, "after the Australian and British are all buried the Turks should all be buried, but not until the burial of our men is completed, because if the Turks begin work before we are finished they are certain to confuse Australians with Turks . . ."[28] The envisioned "ANZAC estate" would allocate the entire northwest section of the Gallipoli peninsula for use as an open-air burial ground and leave the bodies where they had fallen. Bean suggested it should also become the official property of the British Commonwealth.

The request had support from several sectors of the British government, including the prime minister, whose letter to A.J. Balfour in the British Foreign Office suggested that:

> . . . the Gallipoli Peninsula, or that part of it where the remains of many British soldiers are buried, should, when the Peace Treaty with Turkey is being dealt with, be placed under British control for all time, so that it may become sacred territory and saved from the desecration which will inevitably result if allowed to remain under the control of the Turks—and which I am informed has already commenced.[29]

The decision of Australia and New Zealand to take the lead in the commemoration of the Allied dead on the Gallipoli peninsula and to place at least the ANZAC section of the peninsula under the jurisdiction of those two countries was agreed to in an Imperial War Conference of 1917 convened by W.J. Massey, the Prime Minister of New Zealand. In a letter dated 10 October 1919 sent from the official secretary of the London branch of the Commonwealth Offices of Australia, the Australians again urged the Commission to negotiate with the Ottoman Empire at the Versailles Peace Treaty for the entire ANZAC war theater to be "permanently isolated as a sacred memorial to those Dominion soldiers who fell during the campaign; that the land shall be so transferred as a burial ground that it shall not be subject to any form of alien control; and that with a view to its proper preservation and freedom of access, it shall be vested in the control of the Imperial War Graves Commission."[30]

Charles Bean and Alan Box (the latter was the Official Secretary to the Commonwealth in London) noted that the locations of at least 50 percent of all the Australian troops who were killed on the peninsula were unknown. With the ANZAC estate, Bean noted that the problems of exhuming human remains from the difficult topography of that region, and differentiating whether the remains of the dead were Allied or Ottoman would not need to be done with the ANZAC estate solution. He writes:

> The percentages of known burials is low. The bodies, as yet untraced, of some 8,000 Dominion soldiers—and I fear they will never be traced—rest in the neighborhood of Anzac. I think I have said enough to show that merely to cede the few small corners of the historic positions in which those dead of the occupying forces, whom their comrades were able with much difficulty to bury, lie amidst the ravines and rocky spurs of Anzac, would be quite an insufficient grant to cover the actual burial area."[31]

For Bean and Box, creating only isolated patches of formal burial ground in the ANZAC region violated the basic idea of a cemetery. They also noted in the proposal that the vast tract of land for the ANZAC Estate they hoped to claim as a burial ground for the Allied dead could not be of much value to anyone else and would be difficult to organize as a traditional cemetery because of the challenging topography. A holistic approach which encompassed a much larger swath of land with a clearly designated ownership by the Commonwealth was presented as the most appropriate solution.

> Only a very small portion of Anzac is possible for cultivation. To take small spots of land scattered on ledges and small valley flats and to connect such small spots by mountain tracks passing through the land not belonging to either the cemeteries or to the Commission but to an alien nation, leaving out of consideration the majority of the dead whose bones do not rest in those original burial plots, would, I submit, be a negation of the whole idea of cemetery treatment.[32]

In those first steps taken to identifying the Allied dead, Bean encountered immense logistical challenges. The passage of nearly two years since the evacuation, during which time the original wooden markers (crosses), often placed by surviving soldiers for their dead comrades, had disappeared, decreased the possibility of accurate identification.[33] In a letter dated 27 July 1922 written by C. Knowles of the New Zealand Government Office, ultimately the extent of the problem facing the team identifying and categorizing the majority of the Allied dead is apparent.

> With reference to your recent enquiry on the telephone I enclose a schedule shewing the Unknown by nationalities in the cemeteries of the Anzac area. In order to make it complete I have not confined it to the original Anzac areas. I am afraid the number of cases in which the nationality was discoverable was very small.[34]

Knowles' list shows that in the area where he was working only 18 Australian soldiers could be identified accurately as Australian; only 17 New Zealanders could be identified as New Zealanders. The remaining bodies found buried in and around the ANZAC area were just assumed to be one of those two nationalities.[35] Some assistance was given by the Turkish War Office to the members of this first team who came to identify the Allied dead in the ANZAC area. Major Zeki Bey, who had commanded an Ottoman battalion that fought against the Allies on 8 August 1915 and in other battles on the peninsula, assisted the Australians during a week in Februrary 1919 by identifying the areas where Australian soldiers may have been buried.[36] As described by the Australian staff, Zeki Bey "was an intelligent brave little officer who volunteered from the Staff to command a Battalion and his help was most valuable to us."[37] Several moments of the team's collaboration with Zeki Bey were captured by the photographer for the Historical Mission, Hubert Wilkins (Figure 4.2).

The ownership of the ANZAC region of the peninsula has been a contested heritage issue between Turkey and Australia and New Zealand ever since the early days of the GRU's efforts to identify, exhume, and rebury the Commonwealth dead into the ANZAC region and Cape Helles. The sections of the 1923 Treaty of

Figure 4.2 Major Zeki Bey, who commanded an Ottoman battalion that fought against the Allies, assisted the Australians by identifying the areas where Australian soldiers may have been buried. This image shows the Australian Historical Mission at lunch on Hill 60. Zeki Bey, the Turkish officer, is second from left.

Credit: AWM A05258

Lausanne that deal with the Allied graves on the peninsula are quite clear about ownership. Article 128 states that:

> The Turkish Government undertakes to grant the Governments of the British Empire, France and Italy respectively and in perpetuity the land within the Turkish territory in which are situated the graves, cemeteries, ossuaries or memorials of their soldiers and sailors who fell in action or died of wounds, accident or disease, as well as those of prisoners of war and internal civilians who died in captivity.

Article 129 states that "The land to be granted by the Turkish Government will include in particular, as regards the British Empire, the area in the region known as ANZAC."[38]

The CWGC, which followed up the work of the GRU, refrained from insisting that the entire section of the peninsula be designated as a cemetery, and suggested a compromise of sorts.[39] In lieu of the "ANZAC Estate" proposal, where the dead would essentially be buried wherever they fell and a large part of the peninsula would become an immense open-air cemetery, the CWGC decided to exhume some groups of human remains and cluster, or concentrate, them into several different cemeteries scattered throughout the peninsula.

Desecration or desperation?

Today, the collective memory in Turkey, Australia, and New Zealand about the Gallipoli campaign has been shaped by a commemorative narrative which situates the soldiers who came from those three countries as victims manipulated by the larger imperialist forces fighting the war, particularly Britain, France, and Germany.[40] Tales of wartime camaraderie between the ANZACS and the Ottomans—the Johnnies and the Mehmets—are retold and remembered each 25 April (ANZAC Day). Statues and photographs of Ottoman soldiers carrying a wounded ANZAC soldier in a manner recalling the *Pietà*, or giving the last drops of water to a dying Australian, have become iconic images of sacrifice and solidarity in the face of the hardships shared by both sides during the war. While revisionists on both sides have been critical of this idealized discourse of camaraderie of enemies—some referring to it as "Anzakery"—the benefits of the soft diplomacy between Australia and Turkey have outweighed a more intense scrutiny of actual events. As Jenny Macleod has pointed out, "the impulses of these two former enemies remain compatible through identification with two different dates: March 18 and 25 April, and through repeated recourse to the conciliatory sentiments of the words attributed to Mustafa Kemal Atatürk."[41] Macleod is referring to Atatürk's speech quoted earlier in Chapter 2 in which he proclaims that, "there is no difference between the Johnnies and the Mehmets to us where they lie side by side herein this country of ours . . ." This letter, penned by Atatürk, was read to a group of Australian pilgrims in 1934 by the Turkish Minister of the Interior, and excerpts of it are now inscribed on an epigraphic slab at Anzac Cove and other sites around the peninsula.[42]

Typical of this diplomatic discourse from the Turkish side is Kahraman Şakul's *Gallipoli in the Great War and its Heritage*, an exhibition and accompanying volume published in 2016 by the History Department of Şehir University in Istanbul. Here, Şakul suggests that the strong ties that bind Turkey, Australia, and New Zealand in this spirit of amity and cooperation should become an acronym: TANZAC. Under the heading "From ANZAC to TANZAC" he proclaims that "There is no need to make big distinctions between the invaders and the defenders . . . There was no ideological animosity but mutual respect."[43] West, however, has remarked, this "new dialogic mythology" has evolved since the mid-1930s and is based on "the collective memory of the two nations as innocent martyrs."[44]

Absent from these narratives about camaraderie, innocent martyrs and unfortunate victims of larger imperial forces is the discovery that Bean and the members of the GRU made shortly after arriving on the peninsula in 1917: many of the graves of the Allied soldiers on Gallipoli in the Cape Helles area had been desecrated, and to a lesser degree in the ANZAC region. Bean acknowledged that the hardships experienced by the victorious Ottoman soldiers who had remained on the peninsula during the winter months after the withdrawal of the Allied forces had few options for survival, and burned for fuel the improvised wooden crosses that had been made initially by the Allied troops to mark the resting places of their dead comrades. Others, such as Hughes, were not as sympathetic towards the Ottomans' plight. In his first reports during the winter of 1918, he related alarming and critical observations about the condition of the Allied graves on the peninsula (Figure 4.3):[45]

> Practically all the British and French graves at Cape Helles, large numbers elsewhere have been systematically desecrated; this has been done deliberately as it is very clearly spade work and not scratching by jackals, dogs or result of rain. Bones in many cases lying beside graves which have been opened; other cases skeletons lying in open graves. Officers of H.M. Ships landing at Cape Helles November 5th saw a party of Turks filling up graves. Large numbers of graves have been filled up recently, spade marks very clearly showing[46] . . . Not content with rifling the bodies (the Turks) have apparently been playing marbles with the skulls and tip-cat with the bones.[47]

Opinions at the time did differ about the extent and reasons for desecration of the makeshift Allied graves. E.R. Peacock, a British journalist, noted in December 1918 that he found little evidence to substantiate the claims of grave desecration and wrote back to his paper that while land mines made travel and work in the peninsula risky, the real problem was the growing animosity between the Turks and Greeks:

> Re: Cemeteries . . . There have been some gruesome stories of violations, bodies being rooted up, but I have not seen much to justify them—indeed the Turks have put up notices and have enclosed with wire fences (their own iron standards) in some places. The antagonism between the Turks and Greeks is extreme.[48]

Figure 4.3 Australian graves “remade” by the local Turkish residents.

Credit: AWM G02085

With an eye toward preparing the peninsula for family members who could afford to make the journey to see the final resting places of their kin, Hughes initially warned in 1918 that none of the former battlefield landscapes, from the ANZAC area to Cape Helles, was ready to receive visitors as there were bones and skulls exposed everywhere. "These sights would be unspeakably distressing to relations if visited prematurely."[49] By the spring of 1919 Hughes had taken a somewhat more conciliatory approach to the desecration of the Gallipoli graves and reported to Lloyd George on 3 April 1919 that the looting he had seen on the peninsula was largely driven by the poverty of the local inhabitants who did not discriminate between Allied and Ottoman bodies or graves:

> A special mission of Australians is now in Gallipoli inspecting the many thousands of Australian graves there and investigating the systematic desecration which is reported to have been made. Fortunately these reports turn out in the main to be incorrect, but there are abundant evidence that local inhabitants have despoiled a certain number of graves in the hope of discovering money and other valuables. This applies not only to Australians' but to the Turks' also and appears to arise solely from the weakness of Turkish government control. The inhabitants are now returning in great numbers, salvaging the whole of the Anzac for wood, tanks, and other articles of value, and there is no doubt that there will be a recurrence of the disgraceful desecration of graves if special steps are not taken.[50]

On 7 November 1919, Hughes indicated in a letter to Bean that a significant amount of progress in identifying and burying the dead had been made in the ANZAC region of the peninsula thanks to a fire there which had helped to clear much of the underbrush. In his opinion, by the winter of 1919 the horrors of the war had been removed, the landscape sanitized through fire, bones of the dead were now reburied, and the peninsula was ready to receive its first visitors.

> All the bones, for the whole of the Anzac area have been collected, and the place is now in a fit state for people to come here, although I do not think it advisable to advertise the fact. The whole of Anzac was burnt off during the summer and this greatly facilitated matters.[51]

The "other" Allied dead: multi-faith burials, French cemeteries on the Gallipoli peninsula

With its commitment to commemorate every fallen soldier by name and mark on the landscape where (or at least near where) each had died, the Commission also faced the question of what to do with the large number of non-Christian dead who had fought for the Allied forces. There were many discussions about whether the graves of Jews, Hindus, Sikhs, and Muslims who had died fighting on the Gallipoli peninsula with the British Imperial forces required a standardized approach in terms of rank and name. Also pointed out was that some of the design elements and iconographic symbols, such as the cross, had to be rethought for the dead who were of faiths other than Christian.

In Frederic Kenyon's 1918 report "War Graves: How the Cemeteries Abroad Will be Designed," he mentions the challenge of finding an architectural form for a central monument which would be acceptable to the families of soldiers who had died fighting for Britain but who were from faiths and geographies that were not Christian and not Great Britain. While the monument had to be able to "satisfy the religious emotions of as many as possible, it must give no reasonable ground of offence to any."[52] Christians, Jews, Hindus, and Muslims were to be recognized and respected equally for the sacrifices that had been made in battle, and their faiths acknowledged in the different design templates produced by the CWGC. While the need to commemorate the non-Christian dead was considered to be essential, separate burial grounds were initially suggested as the solution for Muslims, Hindus, and all Allied dead of World War One who were of other faiths:

> It will be understood that where our Mohammedan, Hindu, and other non-Christian fellow subjects lie (and care has always been taken to bury them apart) their graves will be treated in accordance with their own religious beliefs and practices, and their own religious symbols will be placed over them.[53]

The British were not the only ones to face the question of how to commemorate the Christian and non-Christian graves in the areas allocated on the peninsula to them for burials. This was also an issue for the French government as they had a significant number of Senegalese and North African Zouaves who had died or had been lost at sea in different battles during the campaign. The French followed a somewhat different approach to burial and commemoration than the British. The French War Memorial was the only cemetery complex on the peninsula created by the Allied nations during the years 1918–1926 that was not part of the CWGC's commemorative efforts during the immediate post-war period. Today the well-tended cemetery with its lavender and rosemary bushes is home to more than 2,000 individual graves, and large white-washed ossuaries hold the others (Figure 4.4). When the commemoration of the Allied losses in the Gallipoli campaign is considered, France has been much less interested than Britain and her former Dominions in remembering her past on the peninsula. This was noticed soon after the war ended by Pemberton, who was working for the CWGC on the peninsula in 1926 and wrote that "The French and the Germans have not deemed it necessary to take such pains as the Anglo Saxons in the commemoration of those of their compatriots who fell on the field of battle."[54]

While Pemberton's observation is not entirely correct, the French, by comparison, had taken far less initiative than the Australians and British in commemorative projects on the peninsula and in the Gallipoli campaign in general. Dutton has insinuated that the lack of interest among French historians in the Gallipoli campaign is intentional and the topic has been a "victim of studied

Figure 4.4 The French cemetery at Morto Bay today.
Credit: Archive of L. Thys-Şenocak

neglect." He notes that only one monument in France—in Marseilles—is dedicated to the French who died in Gallipoli; some recognition of French participation in this theater of World War One was made in the 1930s when a Paris street in the 17th arrondissement was named the Rue des Dardanelles.[55] This is rather suprising as when compared to the British and Irish who had lost more than 22,000 troops, the Australians 7,000, and the New Zealanders 2,700, at least 10,000 of the French who served in Gallipoli did not survive. Six hundred men had been lost at sea in a single incident, when the French battleship *Bouvet* sank in the Dardanelles on 18 March 1915. Many of those who fought and died for France in Gallipoli were from the colonial Senegalese and Zouave regiments who participated in the diversionary landing that the French navy made at Kumkale on the Asian shore of the Dardanelles on 25 April 1915, and in the many land battles fought throughout the summer months in that year (Figure 4.5).

Part of the explanation for the comparative lack of interest the French have shown in creating a narrative about their role in the Gallipoli campaign is due to the resentment they had towards the British commanding officer, General Sir Ian Hamilton. Following the orders of Englishmen had been disastrous and, according to the French, the Gallipoli campaign had diverted troops from the Western Front where they were needed to protect France itself.[56] The British Ambassador in Paris, Lord Bertie of Thame, conveyed in 1915 some of the French sentiments about the British-led campaign:

> The French are very sore on the subject of the Dardanelles Expedition . . . The Dardanelles Expedition is regarded by ordinary Frenchmen as undertaken in the interests of Russian materially, and of England politically, to secure her position in India and Egypt.[57]

However, as Eleanor van Heyningen and others have pointed out, at the beginning of the campaign the French had not been dragged into this particular theater of war; they had, in fact, shown a great deal of enthusiasm in participating in the naval offensive against the Ottomans, as there was the prospect of claiming territorial rewards after a victory. Due to the mapping of the peninsula a few years before World War One, and the surveying that French engineers had done in the late eighteenth and nineteenth centuries for the Ottoman sultan to improve fortifications along the Bosphorus and the Dardanelles, they also had acquired a sophisticated knowledge and interest in the Gallipoli peninsula decades before the war broke out.[58]

During the Gallipoli campaign the French had shared the south end of the peninsula with the British and were stationed for almost nine months in and around the Ottoman fortress of Seddülbahir at Cape Helles. They therefore had a sizeable number of dead to account for and bury. Unlike the British, the French held the "traditional view that an ossuary was the proper receptacle for a soldier's bones and in any case cheaper than providing separate graves."[59]

Figure 4.5 Post-war visit to the French Galinier cemetery which was located within the fortress of Seddülbahir.

The use of ossuaries for the dead is a common practice in France, Germany, and Italy and reflected the stronger religious sentiment about burial and commemoration that was prevalent in France at this time, particularly among French Catholics. Ossuaries formed part of the familiar spatial repertoire in many churches in France; at Gallipoli they provided a focus and tangible site for prayer and pilgrimage.[60]

While the battles were still being fought on the Gallipoli peninsula the French provisionally buried their dead in five different cemeteries; four of these were located on the peninsula. Those who had been evacuated and died in the hospitals of Mudros and Alexandria were buried in the cemeteries adjacent to those sites. The Galinier Cemetery was located in the fortress of Seddülbahir itself; the 1st Division cemetery was at the base of Eski Hisarlık in Morto Bay, where the Turkish Çanakkale Martyrs Memorial is now located; this site was also where the majority of the colonial troops were buried. The dead of the 2nd Division were buried to the north of Morto Bay; the cemetery for the ambulance personnel was located in the Ottoman fortress of Kilitbahir.

The main French cemetery for the World War One dead is today two kilometres north of Seddülbahir, at the landing point on Morto Bay referred to in the war as S Beach. In addition to this main cemetery at the southern end of the peninsula, during the Crimean War (an earlier conflict in the region in which the French had fought along with the Ottomans between 1853 and 1856) the French had established a cemetery, and a small community, along with the British, in the actual town of Gallipoli north of the peninsula. While the British dead from that war had been removed from Gallipoli and reburied in the Haydarpaşa cemetery in Istanbul, the French cemetery is still located in the town of Gallipoli and has an ossuary for 5,000 soldiers who perished in the Crimean War as well as graves for some members from the French consulate who lived in Gallipoli in the late nineteenth century. This cemetery is also the location of a monument honoring eleven Senegalese soldiers who died after World War One during the years of the Allied occupation of the peninsula. Individual names are not provided on the monument and it is written only that "Ici, alentour reposent 11 soldats sénégalais morts pour la patrie française 1919–1923."

In contrast to the British policies, the French government did allow for repatriation of the remains of the dead, and many relatives arranged for this in 1922. But for those families who did not, or could not, arrange for repatriation of the remains of their loved ones, the French government buried these in one of the four temporary French cemeteries on the peninsula, and then reburied those at Seddülbahir in 1923 after the Treaty of Lausanne. Finally these remains were exhumed and reburied a third time at the French National Cemetery and Memorial, when the French government was allocated 28,000 square meters of land at Morto Bay. This cemetery was officially inaugurated on 9 June 1930 by Gourard, the French General who had served on the peninsula during World War One (Figure 4.6).[61]

This large cemetery at Morto Bay contains 2,235 individual graves marked by black metal crosses bound together with small diamond-shaped plaques that give

Figure 4.6 The French National Cemetery opened on 9 June 1930 and all French soldiers who had been buried in different cemeteries on the peninsula were to have been exhumed and reburied a third time at the French National Cemetery and Memorial on 28,000 square meters of land at Morto Bay.

Credit: Archive of L. Thys-Şenocak

the ID number and name of the soldier as well as the inscription "mort pour La France" (Figure 4.7). The individual graves are arranged according to rank; non-Christian and Christian soldiers, Senegalese and Zouaves were buried with all their other comrades. Non-Christian burials are marked with only the vertical section of a cross. There are also several collections of ossuaries in the grounds of the cemetery: the Kilitbahir ossuary with the remains of 22 soldiers who were first buried in that smaller temporary cemetery at the Ottoman fortress, and a large ossuary complex consisting of four sarcophagi. Finally there is a 15-meter-high white obelisk/ossuary which contains the remaining burials. Embedded into this monument are the plaques from many of the former temporary cemeteries built on the peninsula by the French; along with the bones, these plaques were gathered for display on the monument along with stones from several graves and commemorative plaques belonging to different regiments. The inscription on the obelisk reads, "Ave Gallis Immortalis" (Hail, Immortal France) and a commemorative verse, written by Victor Hugo, sits below this line.

Construction challenges

In addition to determining how best to commemorate the different faiths and nationalities on the peninsula, several other logistical problems faced the commemorative projects of the Allies on the Gallipoli peninsula. The cemeteries and commemorative monument projects undertaken by the IWGC required a significant amount, and a particular type, of labor. This need for labor on the peninsula, particularly for qualified stonemasons, required the recruitment of workers with a variety of skills. Hughes's letter to Bean, written in the spring of 1919 from his office at Kilye, attests to the changing demographics and diverse ethnic groups that were coming to find work on the CWGC cemetery projects:[62]

> . . . I have 50 native labourers . . . They can not speak English but my mixture of Arabic, Turkish, and Greek seems to impress them and the fact I'm an Australian is better still. . . . From the way refugees are coming back here there should be no trouble in obtaining laborers . . .[63]

Hughes's work was interrupted for almost a year during the Turkish War of Independence but resumed in the winter of 1923 when a new pool of labor, this time with Italians and Russians fleeing the Bolsheviks, was recruited after the local Greeks had fled the peninsula for the second time since the beginning of World War One.

As Pemberton observed,

> The monument builders, however, were a godsend to the motley bands of people who gathered on the Peninsula, many of them the flotsam and jetsam

Figure 4.7 An individual grave at the French cemetery at Morto Bay.

Credit: Archive of L. Thys-Şenocak

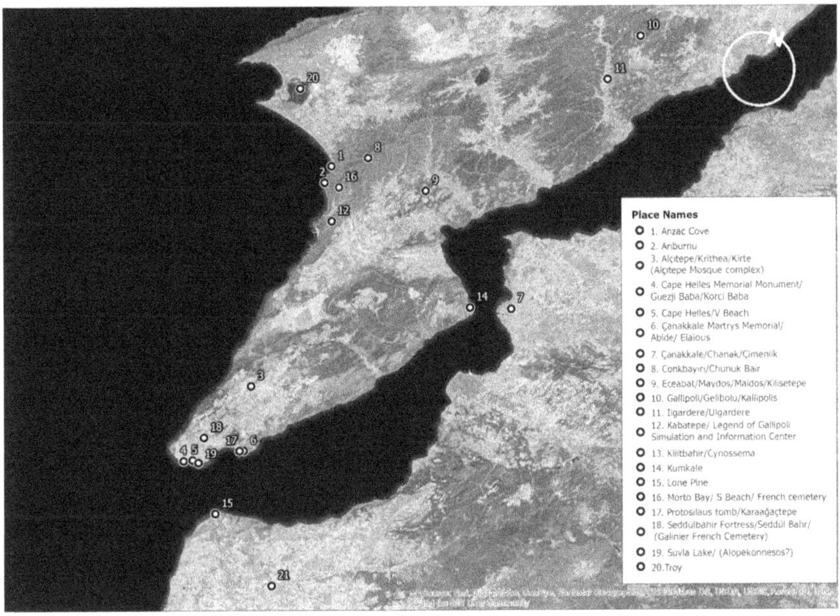

Figure 4.8 Map of the peninsula showing Ilgardere (Ulgardere), and several locations mentioned in this chapter

Credit: Google Earth/Elifgül Doğan

> of a war-devastated world. Among the refugee Russians there were men of birth and education, but they, along with all the rest of the temporary inhabitants were glad to find the means to live.[64]

Most of the roads that were built along the peninsula, many of which continue to be used today, were constructed by the CWGC and its employees. Some of that infrastructure has disappeared, however, including an aerial monorail and light gauge railway that was used to transport the quarried stones of Ilgardere and other supplies across the last 2.5 km of inland gullies and ravines to the cemeteries along the coast (Figure 4.8).

The Allied Memorials

According to the guest register at the Cape Helles Memorial:

> By the Treaty of Lausanne, the Cape Helles Memorial is established in perpetuity to overlook the land and the seas on which, for 3,000 years, Europe and Asia have fought, and on which 12 years ago, the fiercest conflict of all began and ended. It is in no disparagement of the stubborn and successful valour of the Turks that its inscriptions recall the self-sacrifice of

> men from Western Europe, from Australia and New Zealand, from Asia and from Africa, who fought under the British flag. It is a memorial to the dead of an unsuccessful but supremely glorious campaign.

McQuilton, Winter, and others have written extensively about the significance of war memorials, particularly those constructed to remember the fallen of World War One.[65] As mentioned in the first chapter McQuilton states that because so many Australian and New Zealanders lost their lives on the peninsula, "There is little doubt that most Australian and New Zealanders see ANZAC Cove on the Gallipoli peninsula as 'theirs'."[66] As early as 1919 the Australian government had petitioned Britain during post-WWI negotiations with the Ottomans to claim ownership of the first landing area on Arıburnu.[67] Lord Curzon had agreed and posed the rhetorical question: "who had a greater right to be in Gallipoli than the British, with their 25,000 graves on the peninsula?"[68] McQuilton has noted that, "For years, the Allies' memorials dominated the peninsula from Cape Helles to Lone Pine and Chunuk Bair, reinforcing that this was 'ANZAC' space."[69] He and Unwalla have also observed that commemorative sculpture, monuments, and various markers of memory constructed by the Turks are increasingly challenging this sense of ownership of the ANZAC space today.[70]

The construction of commemorative monuments by the Ottoman government did begin in the first months of World War One and will be discussed in more detail in the following chapter, but since the construction of the ANZAC commemorative site in 1999, (300 meters north of Arıburnu and now the site of the annual Dawn Ceremony), ANZAC space on the peninsula is equated with the beaches of Arıburnu, and all other regions of the peninsula are of secondary importance.

Some of the visitors from Australia and New Zealand who come to participate in the annual dawn ceremony each 25 April do not even visit the southern section of the peninsula, even though ANZAC troops also participated in the invasion at Cape Helles, and some of their compatriots who died on other parts of the peninsula are commemorated at this monument. As one of the main guides to the peninsula notes in a perfunctory manner, "Although the ANZACs shared in the Helles project, the Australians required their own memorial at Lone Pine, the New Zealanders (who have no names on this memorial) theirs at Hill 60, Twelve Tree Copse, and Chunuk Bair." The usual post-dawn Gallipoli itinerary for the Australian or New Zealander is described as follows:

> Once the Dawn Service is over, participants have time to wander up the hill to the Australian ceremony held at Lone Pine Cemetery and the New Zealand ceremony held at Chunuk Bair. Most take Artillery Road and stop along the way to visit Shell Green Cemetery where the evacuating ANZACs played a game of cricket to dupe the Turks into thinking it was business as usual behind ANZAC lines. Some wander into the bushes or climb the cliffs and ridges. The opportunity to interact with the Gallipoli landscape heightens

many of the visitors' emotional response to the event, and gives the experience greater depth of meaning.[71]

The landscape and commemorative spaces that a typical ANZAC tourist experiences are thus frequently limited to the Arıburnu area and the ridges behind it. As Macleod and many historians of Gallipoli have pointed out, the number of ANZACs who died on the peninsula, however tragic, is far less than the number from Great Britain, including the many Irish who lost their lives on the peninsula.[72] There has been criticism about the "ANZAC-centric" aspects of commemorating the loss of life at Gallipoli, with the argument put forward that there were other tragic days during the battles fought on the peninsula where the loss of life was just as great, if not more than at Arıburnu. The identity of the dead from other Allied countries or member states of the British Commonwealth, while greater in number than the ANZACs, are somehow portrayed as less important and receive relatively less commemorative time and space.[73] As Macleod has noted, "The Britishness of the Gallipoli defeat notwithstanding, to write about the memory of the Gallipoli campaign in Great Britain and Ireland is, in large measure, to write about the commemoration of the Anzacs rather than the more numerous participants from the British Isles."[74] Australia and New Zealand's primary interest in the northern section of the peninsula and the beaches of Arıburnu is understandable as their main offensive was there and most of the soldiers from these countries died in this section of the peninsula, but the contestation over commemorative space and the monumental architecture that shapes this space extends beyond McQuilton's perception of a schism between Turks and includes different views on how to weigh the sacrifices made by the different participants from former Allied countries.

The Cape Helles Memorial

The largest commemorative monument undertaken on the peninsula by the British is the Cape Helles Memorial, and its construction process attests to the tensions confronted by the British as they strove to commemorate their dead on a landscape which they had neither conquered nor with which they had a strong religious or cultural affinity. A passage in the register of the Cape Helles Memorial reveals the ambiguity and contradiction upon which the entire project was built: "It is a memorial to the dead of an unsuccessful but a supremely glorious campaign."[75]

As Fuchs has remarked, "Owing to the decision not to repatriate but to bury the dead in the vicinity of the battlefields, the cemeteries became more than sites of mourning; they became historical documents that recorded some of the realities of the war."[76] This is certainly true of the Cape Helles Memorial, which is a record in stone of the names of battleships, battle cruisers, divisions, brigades, lines of communication units, names of the British, Australian, and Indian troops who died but whose graves were never found, and the troops who were buried at sea. There are so many names listed on the memorial that one popular guide book to the Gallipoli peninsula describes searching for a name on the Cape Helles Memorial like "looking for a needle in a haystack."[77]

The architect who was chosen to design the Cape Helles Memorial was Sir John Burnet (1857–1938), a Scot who had been a lieutenant in Egypt and had worked for the CWGC in Palestine. Of the five major memorials built by the Allies on the Gallipoli peninsula, Burnet designed three: the Lone Pine Memorial, Twelve Tree Copse, and the Cape Helles Memorial, the latter being the architect's most impressive monument. Hurst Seager from New Zealand was responsible for the remaining two monuments commissioned at this time for Chunuk Bair (Conkbayırı) and Hill 60. Burnet was also responsible for the architectural and landscaping designs for 31 of the CWGC cemeteries. As one of the architects employed by the Commission, he was expected to work with the accepted design principles, iconography and aesthetics that the Commission had established for monuments and cemeteries dedicated to the memory of Commonwealth soldiers everywhere. Within those parameters, Burnet proved able to adapt his designs for monuments, tombstones, and landscaping to the requirements of the topography of the peninsula as well as its cultural context.[78]

Sir John Burnet was a well-established architect with offices in London and Glasgow; he worked both in Gallipoli and Palestine with two partners, Thomas Tait and Lieutenant David Raeside of the Royal Engineers, the latter who turned Burnet's sketches into workable drawings.[79] Burnet had his share of critics, particularly because he was slow to complete the projects to which he had been assigned, and wasn't in the habit of staying within his budget. According to Hughes, Burnet was "inclined to dally."[80]

In a confidential letter dated 15 October 1919 to Sir Frederic Kenyon at the British Museum, the Major-General wrote:

> . . . Burnet, with the assistance of Raeside has made certain sketch designs which undoubtedly will work out at a higher cost than that laid down in the unit cost agreed to. Burnett is working very slowly and evidently regards this as a ten years' business. The result is, as you will understand, that the whole of our work in Palestine is held up.[81]

Burnet's report to the Imperial War Graves Commission dated 30 June 1919 states that he was accompanied to the Gallipoli peninsula by two other assistants: Lieut. Keesing and Major Everett. Once on the peninsula they met with Bigg-Withers and Hughes, who were still there working to identify graves. One of the interesting points mentioned in this report is that while touring the peninsula Hughes gave Burnet a Turkish survey map which indicated several places where makeshift cemeteries had been created by Allied and Ottoman troops during the war. Burnet writes that "Lieut. Hughes reported that the Turkish surveys were more correct than the British maps."[82]

The Australians had also discovered that the Ottomans had excellent sets of maps of the Gallipoli peninsula. Among these must have been the Şevki Pasha maps, as is apparent in the anecdotal report sent by Hughes to Bean about the former's experience in the Turkish War Office in Istanbul:

> I went down to the Turkish War Office at Stamboul and saw the old Bey, who was rather loathe to part with the maps, and who said: "You British were at Suvla. How is it that you did not make maps? Had you not the time?" By then the maps were in my possession, so I replied: "Yes; but I knew that some day a British officer would stand in the Turkish War Office and demand any maps he required." He smiled somewhat thinly; but I could see that he did not like it.[83]

Due to the presence of live mines in certain parts of the peninsula, Burnet and his team were unable to visit some locations on the peninsula such as Suvla Bay and the northwest, but learned about these areas from aerial photographs. Burnet was tasked with determining whether the commissioned Cape Helles Memorial would be better suited in the ANZAC, Suvla, or Helles region of the peninsula. From the areas that he was able to visit, Burnet observed that the land in the ANZAC region had "traces of extensive landslides which, though covered at the time of our visit with a rough scrub, indicate in my opinion unreliable and insecure ground unsuitable as foundations for permanent monuments of any size or weight."[84]

Burnet was also unsupportive of the Australian proposal for the above-mentioned "ANZAC Estate" and suggested a more practical and manageable program for burials in this region and "that the number of cemeteries be reduced by a further concentration of burials to those cemeteries which are on safe ground or can be adequately protected from all risk of damage by surface water or other forces."[85] According to Burnet, the political difficulties of acquiring such a large tract of land from the Ottoman government to create a massive ANZAC estate were daunting, so Burnet focused on the topographical challenges instead. The "ANZAC Estate," he warned, was an unrealistic and unsustainable proposal.

> . . . had the district been an ordinary one, or even like Helles or Suvla Bay, I think nothing better could have been done; but the site being what it is it could not be left in its present condition . . . and even if the whole ground were to be very carefully laid out and planted, and the roads or paths well and thoroughly made, their upkeep, in my opinion, would involve careful annual inspection and considerable labor . . .[86]

Other suggestions made at this time by Burnet in the same report appear to be motivated more by pragmatic concerns of erosion and the challenging topography of Gallipoli rather than its aesthetics, and included some general recommendations concerning how to wall all the cemeteries "with finely built walling at least 8 feet high designed as far as possible in simple level lines culminating in a raised portion on the highest side of the cemetery. . ." These walls would serve "as a protection against the shifting nature of the soil, they should be surrounded external to the walling by a belt of trees, say Eucalyptus, Pepper or other evergreen, and internally with Cypress trees."

Having visited the peninsula first in the spring of 1919, by 5 October 1922 Burnet had submitted to the Commission his design and cost estimates for the Cape Helles monument and the project was ready to go to tender.[87] Burnet's design for the Helles memorial was simple and functional: "square on plan, is 25 feet at its base, tapering 10'6" at the top, and rises in the form of an obelisk to a height of 99'6" above the platform on which it stands" (Figure 4.9).[88]

In Europe of the 1920–30s, and on the Gallipoli peninsula, the obelisk was a less controversial and a standard iconographic choice for commemorative architecture. While Christian iconography was used in many English war memorials for its "accessible and palliative language" [89] and the cross features prominently in commemorative monuments in the Allied cemeteries of Europe, it was not considered appropriate for Gallipoli. Figurative sculpture was also considered to be problematic because of the large number of non-Christians whose names were commemorated on the walls of the monument, and because the monument was to be located in a Muslim country.[90] Further, Burnet was concerned that using a monumental and freestanding Christian cross would be too prominent as a religious symbol in a Muslim country and might invite vandalism.[91]

Burnet suggested that an obelisk without an inscribed cross was a safer choice and one which was part of both the British and the French stylistic repertoire for commemorative war monuments. The architect's choice of this particular form

Figure 4.9 Sir John Burnet's design for the Cape Helles monument.

Credit: Archive of L. Thys-Şenocak

was not unique. As Jay Winter has pointed out, "Some English communities carried on the old Puritan war against graven images, preferring obelisks to crosses. Obelisks abound in France, too, in part because they were cheaper to build than figurative art."[92] Yılmaz and others have noted that the choice of an obelisk, a phallic form, was also a highly gendered iconographic statement as it conveys ideals of virility and "embodies masculine ideals of conquest and domination that are typically considered to be prominent features of monuments representing war in a heroic light."[93] While the High Commission disagreed with Burnet's sentiments concerning the problems of using a freestanding Christian cross in the Commonwealth War Grave cemeteries in Jerusalem and Gaza, they did opt eventually to take Burnet's recommendation for the Gallipoli peninsula, and did not insist on a Christian cross at Cape Helles.[94] As one journal at the time reported, it is "interesting to note that in Mohammedan countries religious approaches are respected and the symbol of the cross is not given a more than necessary prominence."[95]

Research conducted about the geological past of the peninsula had raised concerns for Burnet about the impact of earthquakes on the stability of the monument, so precautionary measures were taken and the memorial was "designed with an interior structure of reinforced concrete secured to a reinforced concrete base, about 38'6" square and 6'6" in thickness." This concrete core was then to be faced with stone quarried from Ilgardere, "joggled with cement and sand joggles with headers bonded into the concrete, the quoins to be cramped with strong galvanized iron cramps."[96]

The design for the Cape Helles monument efficiently made use of existing materials that had been left behind by the Allied forces when they evacuated, including the railway ties of the Decauville railway, a light-gauge rail line which had been installed by the Allied troops to move ordnance around the peninsula. The salvaged rails were embedded into the foundations of the monument, creating an iron "raft" that was hoped would provide resilience in case of an earthquake.[97]

Suggestions for the material, location, and even the font size to be used for the inscribed names of the fallen on the panels along the interior courtyard of the monument were all discussed, and the cost of the monument was estimated to be £54,820, with £25,280 of that total allocated for the actual construction. The inscriptions, estimated at £1 per name, and the imported Hopton Wood grave markers were the second most costly items in the budget, at £16,290.[98] Burnet justified the higher cost of these individual imported grave stones because he felt that the individual white crosses were not aesthetically pleasing (Figures 4.10 and 4.11):

> As present[ly] arranged the burials are so close that the present white crosses touch and in some cases appear to interlace with one another, actually giving the appearance of fences. This effect is destructive of all quiet and offers a very violent contrast between the fallen who have been identified and those who have not.[99]

Figure 4.10 An example of an individual tombstone from the CWGC cemeteries.
Credit: Archive of L. Thys-Şenocak

Figure 4.11 Burnet eventually decided not to use the crosses seen here as individual markers for the graves in the CWGC cemeteries.

Credit: Courtesy of the CWGC archives

Rather, to commemorate the Commonwealth dead, Burnet suggested and sketched individual grave markers. These would be "simple sloped stone about 14" × 12" and standing about 6" to 8" above the level of the ground, and would bear the number and name of the individual" (Figure 4.12).[100] The stone quarried from Ilgardere was selected because it was thought it would weather well in the harsh winters of the peninsula, and also because its lighter color "would always stand out distinctly against the darker background of the hills."[101] The quarry at Ilgardere had not been used for almost fifty years and the CWGC masons assumed from samples that they had taken from the walls at Troy that the ancient inhabitants of Priam's city may have used this quarry on the peninsula across the Dardanelles. It was believed that by using stone similar to that at Troy both the monument and the headstones would stand the test of time, and there was the added prestige that the stone was of an origin and composition similar to that used by the Ancient heroes of Homer's epics (Figure 4.13).[102]

The location of the Cape Helles monument on the peninsula had been a topic of much debate, and there had been several suggestions from different members of the CWGC about where to situate this major commemorative monument of the Allied forces on the ridges above the ANZAC Cove and Tekke Burnu. From Burnet's report, and particularly the notes in the margins of these reports where he scribbled "monument useless at ANZAC," it is clear that he was against locating the memorial in the ANZAC region because of the unsuitable topography and lack of visual impact from the sea. There was, he claimed, "no secure foundation in this immediate front for any central or crowning feature of the size

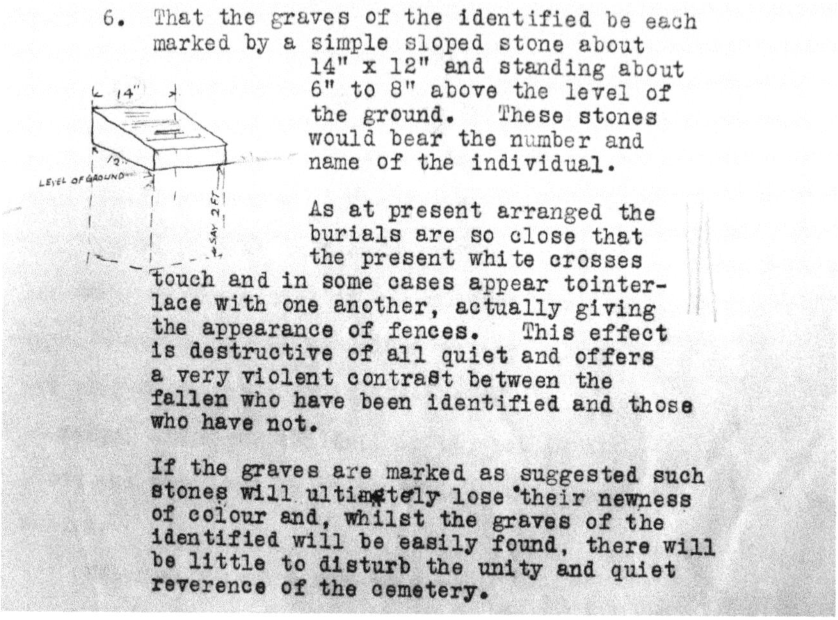

6. That the graves of the identified be each marked by a simple sloped stone about 14" x 12" and standing about 6" to 8" above the level of the ground. These stones would bear the number and name of the individual.

As at present arranged the burials are so close that the present white crosses touch and in some cases appear tointer-lace with one another, actually giving the appearance of fences. This effect is destructive of all quiet and offers a very violent contrast between the fallen who have been identified and those who have not.

If the graves are marked as suggested such stones will ultimately lose their newness of colour and, whilst the graves of the identified will be easily found, there will be little to disturb the unity and quiet reverence of the cemetery.

Figure 4.12 Burnet's sketch for the individual tombstones which he intended to be slightly inclined as he felt these would be more durable and less prone to vandalism.

Credit: Courtesy of the CWGC archives

and scale necessary and, perhaps most important of all, even were any such monument design obtainable, its significance would be overlooked as it is not on that side of the peninsula that the traffic of the sea lies, and it would not be visible even in the distance from this traffic route."[103]

In the same report Burnet suggested that the more appropriate and visible choices would be:

> "on the great rocky cliff (some 65 feet high) forming the east side of Cape Helles, or within the ruins of the old castle of Sidd-il-Bahr on the other side of the Bay."

The final decision about the location of the Cape Helles monument was made two years after Burnet's initial visit in 1919, and only after a second committee had assessed the possible locations available. Seven different areas of the region, including the islands of Lemnos and Imbros, were evaluated by the committee as possible sites. Following Burnet's advice, the committee's first decision was that a major monument in the ANZAC region would be "far off the track taken by any vessels, other than local sailing craft, and no Memorial placed here could be seen, except through a telescope, by steamers sailing up the Dardanelles from Australia, New Zealand, India, Egypt, or Great Britain."[104] Achi Baba, the highest point on the western side of the peninsula at 718 feet above sea level, was also deemed unsuitable because it was "very disappointing when seen again from the sea."[105] The search thereafter focused on the southern end of the peninsula, closer to the entrance to the Dardanelles.

Hunter-Weston and his team, sailing through the Dardanelles on the HMS *Byron*, sketched out the sight lines of each location in the Cape Helles area (Figures 4.14(a) and (b)). Of four possible sites at the entrance to the Dardanelles—Tekke Burnu, Beyaz Tepe/Hunter Weston Hill, Guezji Baba, and Sedd el Bahr Hill—the fourth choice was eliminated as it was thought that it would compete for visual impact with the Ottoman fortress and the village. Then, the interior of the Seddülbahir fortress was eliminated as a possible site for the monument because it was too closely associated with the French, especially as they had buried their own dead in the fortress itself prior to staking a claim to burial grounds at Morto Bay. Guezji Baba, on the other hand, was considered a more viable option since it had been "consistently in British possession throughout the whole period of our occupation."[106] The decision therefore was that the Cape Helles Memorial would be sited at "Guezji Baba" on a point that was 150 feet

Figure 4.13 The quarry at Ilgardere was thought to have been used in ancient times for the masonry walls at Troy and was mined again by the CWGC for their commemorative architecture.

Credit: Courtesy of the CWGC archives

above sea level and 500 yards northeast of the Helles Lighthouse. The monument would stand 60 feet high, be able to withstand earthquakes, and be very simple and enduring.[107] Anticipating acts of vandalism and theft, the committee concluded that there would be "nothing about it (such as bronze plates or very beautiful and costly stone) that is likely to be removed in future ages."[108] Finally, in addition to the visual impact of a monument at Cape Helles, the committee felt that Cape Helles would be the most appropriate place for the monument because "It was here that the original British landing was made; it was here that the greatest amount of Turkish territory was occupied; and it was from here that the last troops were evacuated."[109]

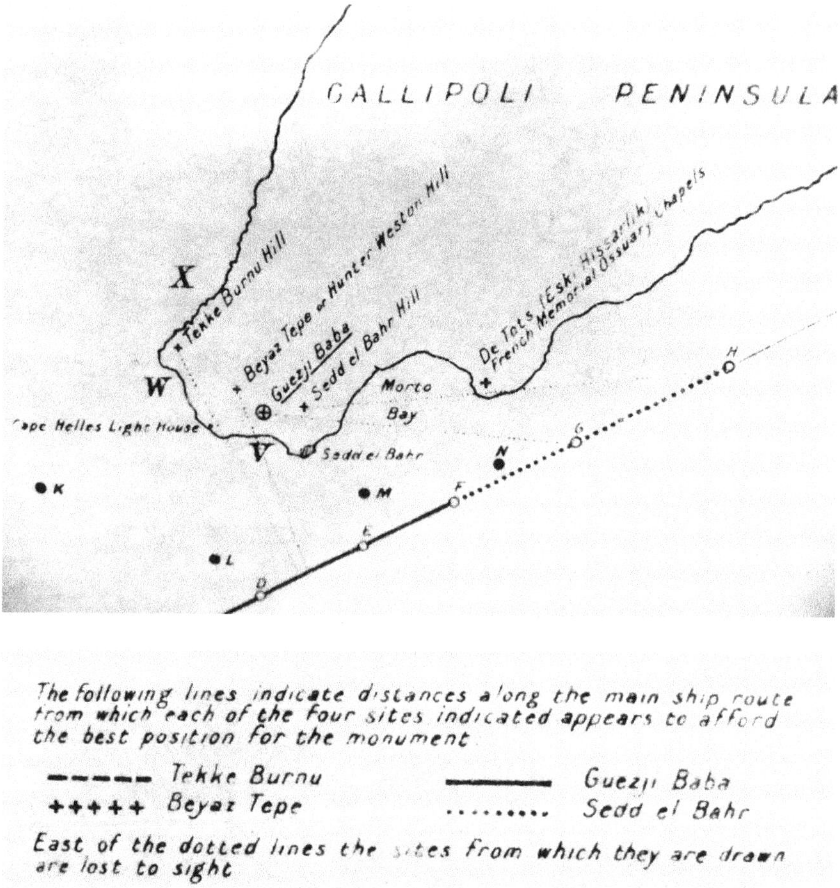

Figures 4.14(a) and (b) Sketch of the site lines in the 14 June 1921 report about the best location for the Cape Helles monument.

Credit: Courtesy of the CWGC archives

The authors of the report expressed hope that the Cape Helles Memorial would be more than a commemorative site for relatives to visit during their personal pilgrimages. Rather, it was to become a prominent mark on the landscape for a much wider audience; it had to be "designed to arrest the attention of those traveling by sea. It should consequently be placed in such a position that it shall be seen by all ships approaching and passing up the Dardanelles on the ordinary sailing courses."[110] The newspapers and press releases in the 1920s further indicate that the Cape Helles Memorial was never intended to be a modest structure. In "The Monument on Cape Helles: A Landmark for All Time" the enthusiastic writer proclaims his hopes that "In time it may acquire in the East, this memorial to our missing dead, a legendary fame such as was given to the Colossus of Rhodes, which was of the same height" (Figures 4.15(a) (b)).[111]

Like all the cemeteries on the Gallipoli peninsula designed by Burnet, the entire monument was intended to give the appearance, from the outside, of a secluded garden. Surrounded by walls and, in the case of the Cape Helles monument, elevated on a platform, "the cemetery would become a quiet and reverent 'God's Acre' reverently enclosed and symbolic of that unity of action and absence of individuality which characterizes the War."[112] The Cape Helles monument's strong division of space allows for more contemplative collective mourning but the design also has, as Winter has pointed out, a practical side: "to protect the monument from accidental damage through contact with passer-bys, or even from the attentions of grazing animals."[113] Ultimately the commemorative processes undertaken by the CWGC and their head architect fulfill the prediction that Charles Bean made when he first dreamed of what the Gallipoli peninsula would look like once the Commission had completed its work:

> The day is far off (but it will come) when splendid mausolea will be raised over these heroic dead. And one foresees the time when steamers will bear up the Aegean pilgrims come to honour at the resting places of friends and kindred, and to move over the charred battlefields of Turkey.[114]

Post-war pilgrimage, battlefield tourism

As predicted, pilgrims and battlefield tourists did come to Gallipoli. By the late 1920s, the vacated and recently manicured and "prepared" war zones of the Western Front—along with the battlefields of the Gallipoli peninsula—began to accept civilian visitors and the phenomenon which David Lloyd refers to as "battlefield tourism" began to flourish.[115] This was partially the result of concurrent developments in the steamship and railway industries in Europe, but was coupled with a surge of spiritualism in Europe and Britain after so many deaths in World War One. Chris Ryan classifies wartime battlefields into "hot" and "cold" tourism, noting that the high point of "hot" war tourism was the period immediately after World War One: "The scale of the devastation, the nature of the warfare, together with the economic, social and technological changes or figurations, meant that an unprecedented number of people could visit the sites of conflict."[116] Some came

Figures 4.15(a) and (b) The opening ceremony at the Cape Helles monument.
Credit: Courtesy of the CWGC archives

to the peninsula hoping to make small fortunes collecting leftover ordnance, or war memorabilia left on the battlefields or for sale by the locals. But others were grieving relatives, legitimate pilgrims seeking solace in the battlefields—the relatives, and comrades of those who had perished in battle. As Lloyd points out, quite often, "The language of the sacred and the act of pilgrimage both infused and were in conflict with battlefield tourism. The interaction between the two modes of travel was a product of the concurrent development of a tourism industry and the renewal of the practice of pilgrimage in the years prior to and after the war."[117]

Several of these battlefield "pilgrims," like some tourists today, were not welcomed by the representatives of the Turkish government, or locals, and their behavior was seen as far from exemplary. At times it was perceived as irreverent; often it was just rude. Then as today, tourists to the Gallipoli battlefields, and those who organized the battlefield tours, were a mixed bag. As one journalist for the *Daily Express* wrote in 1919, these post-war visitors were "morbid seekers after sensation. Vandals. Ghouls of the Battlefield."[118] Early tour agencies who organized trips to the battlefields of World War One, such as Thomas Cook and Michelin, attempted to instruct tourists, through guidebooks and supervised tours, on proper battlefield behavior and respect for the dead. Thomas Cook, for example, admonished those who signed on to the company's tours that "The battle areas will therefore be visited by the tourist not as a show but as a shrine."[119]

The St. Barnabas Society was one charitable organization set up to help the sincere "pilgrims" of England, New Zealand, and Australia to get to Gallipoli and other battlefield sites and cemeteries of the Mediterranean to visit their dead. It was founded by Reverend Mullineux of New Zealand as a charity for the post-war pilgrims, but in a report to the Imperial War Graves Commission (IWGC), one Barnabas voyage in 1928 received complaints that "the crew had been hastily engaged and were almost all of them inexperienced. The stewards had not been sufficiently careful of the Pilgrims' comfort, the bathing arrangements had not been up to the required standard, and on one occasion the water (for Mr. Mullineux's) own bath polluted by yellow rust and ran yellow!"[120] The crux of the problem, the report concluded, was that "not more than half of the party were genuine Pilgrims. They were mere tourists and determined to get all they could and give nothing."[121] Local guides could be a problem as well. In another report from an earlier St. Barnabas voyage of 1926, Mr. Mullineux accused one member of the Commission's staff on the peninsula, who had been appointed as the pilgrims' Gallipoli-based liaison, of being entirely negligent of his duties because he "came aboard the ship reeling drunk."[122]

The staff of the CWGC were the local representatives of the Commonwealth on the Gallipoli peninsula and were frequently called upon to be diplomatic liaisons during visits by pilgrims, foreign dignitaries, and veterans' groups from Great Britain and Australia; this was in addition to their duties tending to the Commonwealth graves. Britian and the CWGC were perceived to be securely ensconced on the peninsula and received requests from France, Russia, and Egypt to maintain their graves on the peninsula, the town of Gallipoli, and the nearby islands of Tenedos and Lemnos. Much of the CWGC correspondence from the late 1920s through the 1970s reveals the Commission's worries about maintaining an office and official presence on the peninsula, and ensuring that the land granted to them under the Treaty of Lausanne was not used or vandalized by locals. Additionally they had to contend with the grieving families of Commonwealth soldiers, some of whom came to the peninsula with their own commemorative plaques to place by a grave or plans to construct memorials while visiting the peninsula.

One report by a CWGC officer who accompanied the 1928 St. Barnabas pilgrimage mentioned earlier, reveals the kinds of challenges that the CWGC faced in the post-war era of maintaining the standardized rules about Commonwealth burials. While visiting the Pink Farm cemeteries, the officer wrote:

> My attention was drawn to a bronze badge of the Gurkha Rifles which Rule Jones had had fixed on a concrete block placed immediately in front of our block containing a tablet with the Officer's name, Lieutenant Beddy. Mrs. Beddy, his mother, was there and rather flaunted it in my face. I had to tell her that such a thing was not permitted and that I would have to report it. I could not say whether it would be allowed to remain or not. She was highly indignant and threatened to haunt me for the rest of my days if it was interfered with in any way.[123]

Maintaining the cemeteries of the Gallipoli peninsula

From the correspondence between the Turkish government and the CWGC, the local resentment of residents on the peninsula was a growing concern. Graves were occasionally found vandalized; the Turkish government had complained to the British embassy that the only signposts put up by the Commission were in English and these should also be in the new Turkish characters; caretakers of the CWGC graves were accused of restricting local shepherds from grazing their animals on the land that did not belong to the foreign cemeteries. A letter written on 1 October 1928 by the British Embassy in Constantinople to the caretaker of the Gallipoli graves, Mr. Millington, noted that word had reached the embassy that "your people have warned the local inhabitants off ground adjoining the cemeteries but not belonging to our concessions and that for example the shepherds had not been permitted to remain on the pasture nearby."[124]

An undercurrent of hostility and resentment on the part of the Turkish government against the CWGC, and a desire for them to vacate their offices at Kelia on the peninsula itself, is evident in much of the correspondence during the late 1920s and 1930s. For its part, the CWGC recorded several problems with increasing government restrictions on their activities and difficulties with locals including theft, vandalism, graffiti, and even assault of its personnel while tending to the graves on the peninsula. A list of the following grievances was reported by the CWGC during their work on the peninsula:

1 Loss of pumps and tools at Sedd el Bahr.
2 Loss of parts from Motor Cars and Tools from Workshops, Kelia.
3 Breaking into Commission's office and interfering with papers.
4 Stopping the use of Motor Boat at Kolia-Chanak.
5 Small defacement of Cross at Chunuk Bair and rude inscriptions in Turkish.
6 Refusal to grant Jones [*CWGC representative*] a pass and on visit to Troy with other Europeans at Chanack, his being the only one refused.
7 News by telegram on November 23rd from Kelia of deliberate breaking of two Coping Stones at Chunuk Bair and the stabbing of one of our Turkish Watchmen and Gardener in a quarrel with Turks, who were digging up remains for brass and cartridges near the Memorial.[125]

In spite of the CWGC's growing difficulties on the peninsula, in 1947 they agreed to look after the graves of the French troops who had died in the Gallipoli campaigns, as long as this arrangement was not officially documented in the accounts of the Commission and that the Commission was reimbursed the maintenance costs by the French Consul-General. Tending to the French graves on the peninsula and nearby islands had been an ongoing problem for the French since the war's end; they frequently had to rely on the assistance of the British who had more experience negotiating with the Turkish government, using the Treaty of Lausanne as justification for their presence in the region. Animosity towards the foreign graves was not only expressed by the Turks on the Gallipoli peninsula but also felt by the local Greeks residing on the islands who had provided support to the Allied troops during the war. During one early post-war visit in 1922 of French military officers to Mudros Harbor on the Greek island of Limnos (Lemnos), for the purpose of painting the crosses on the graves white and grey, the crew found they were decidedly unwelcome by the local population. A member of the Commission reported the incident to his superior:

> The crew received a hostile reception from the natives of the island, who informed the French that they hoped their intention was to remove all the French graves from the island as they did not wish them there. Efforts to purchase eggs and vegetables on the part of the Officers and crew resulted in complete failure, as the inhabitants refused to sell. One café proprietor refused to serve a French Officer and passed insulting remarks. The above is not very pleasant or interesting reading but I mention it to enable you to understand how the inhabitants of the island dislike the French.[126]

French access to their graves had become even more difficult after the signing of the Montreux Convention because their graves at Morto Bay were designated to be in a military zone. The British, however, could access this area and the graves at Cape Helles by boat without passing through the restricted zone, and agreed, albeit reluctantly, to tend to the French graves. As E.H. Peck from the Foreign Office in Britain reported to the Commission on this informal arrangement:

> I would suggest, if you see no objection, that I should inform the French Embassy here, verbally and unofficially, that the Imperial War Graves Commission have an objection in principle to assuming foreign commitments but that the present unofficial arrangement seems to suit all concerned and might continue undisturbed.[127]

The Commission was not, however, willing to continue to undertake this task for the graves of the White Russians who had fled the Bolsheviks in 1922 and who asked in 1931 for both a regular inspection of their graves in the town of Gallipoli and for the Commission to bear the cost of the upkeep of two cemeteries there. Wary of creating a diplomatic crisis with the Soviet government, the Commission warned that "Colonel Hughes and Captain Millington should be told at once of the Commission's attitude, and warned that anything they may do for the Russians must be entirely unofficial and that they must not give the Turkish government any excuse for informing the Soviet Government that the Commission have taken over charge of the White Russian graves."[128] Mr. Millington of the Commission did end up conducting an inspection of the cemetery and writing up a rather extensive report on the very poor state of the Russian graves; the cemetery "was completely overgrown with weeds, and it would appear no attention has been given for a considerable time" but he noted that "the Commission had no funds which they could devote to the maintenance of these graves, nor was there any prospect of obtaining any for the purpose. The Commission's funds were very strictly limited, and could only be spent on certain specific objects."[129]

For many decades after the war, the CWGC continued to fend off legal claims to the land that it had been allocated under the Treaty of Lausanne for its cemeteries, commemorative monuments, and caretaker spaces. As John Paton from the CWGC wrote in the opening paragraph of his letter to the Consular Department of the Foreign and Commonwealth Office in London: "As you will know the Commission have to keep continual watch over the various activities that occur on the Gallipoli Peninsula to ensure that no encroachment is made on the land allocated to the Commission under the terms of the Lausanne Treaty."[130]

In addition to the aforementioned property claims made by the Greek priest of Lesbos and the grocer at Seddülbahir, claims continued to be made against the property of the CWGC, or contestation over the use of the land allocated for their cemeteries. In 1967 Fevzi Gürses managed to officially register a plot of land within the Arıburnu area; in 1968 Ali Çelikbilek was notified that he had encroached on CWGC land in the same region and that he could no longer farm there. While Mr. Çelikbilek complied with little resistance, Mr. Gürses pursued his case through six different court hearings, basing his claim to the land on a property title deed from 1896 which was in the name of Hafiz Ahmet and of whom he was the heir. As of 7 July 1970 the case had not been resolved; it is uncertain what happened in the end, as the correspondence (marked "confidential") between a J.F. Skinner and a second British lawyer in the consular office on that date, and a few weeks later on 23 July 1970, notes that "it is disappointing to hear that the problems have not been brought to the conclusion we had hoped.

However, it is gratifying to hear that the Ambassador will continue to afford the Commission the same facilities as it has enjoyed in the past. We full(y) appreciate your point that the Turks must not become aware of such an arrangement. Yours sincerely, S.S. Talman."[131]

The struggle to exhume the dead from a battlefield with such a challenging topography, carve out spaces for reburial, design monuments for commemoration, and initiate the ceremonies which have now been performed for more than a century on the Gallipoli peninsula was a far more complicated and contentious process than evidenced by the rounded statistics about the dead with which this chapter began. Pheroze Unwalla posits that the very presence of the commemorative monuments and spaces constructed by the IWGC and their successors, the CWGC, on the Gallipoli peninsula provided the Turks with a model for how to commemorate their dead, but has also been responsible for the failure of the latter to create a cohesive national space of commemoration. The cemeteries of the former enemies have "disrupted, then and now, the Turkish national space of the Gallipoli peninsula, perpetually impeding the desire for spatial homogenization and eliciting further fears of foreign control and occupation . . . consequently a space which is supposed to embody a foundational moment for the Turkish nation is hindered from doing so."[132]

Whether this was the impact of the commemorative projects undertaken by the former Allied forces on the Ottomans during and after World War One, and on the Turks in the decades since, is the topic of the following chapters.

Notes

1 R. Reid, I. McGibbon and S. Midford, "Remembering Gallipoli," in *Anzac Battlefield*, 199.

2 See Treaty Series no. 16 (1923) "Treaty of Peace with Turkey, and other Instruments, Signed at Lausanne on July 24, 1923," (http://treaties.fco.gov.uk/docs/pdf/1923/ts0016-1.pdf.)

3 "The Gallipoli Campaign," Commonwealth War Graves Commission brochure (1915).

4 See https://nzhistory.govt.nz/media/interactive/gallipoli-casualties-country for recent statistics. G. Kingsley Ward and Major Edwin Gibson, *Courage Remembered: The Story behind the Construction and Maintenance of the Commonwealth's Military Cemeteries and Memorials of the Wars 1914–1918 and 1939–1945* (London: HMSO, 1988, reprinted 1995), 144.

5 Thomas W. Laqueur, "Memory and Naming," in *The Great War: The Politics of National Identity*, ed. John R. Gillis (Princeton: Princeton University Press, 1994), 152–3.

6 Laqueur, "Memory and Naming," 155.

7 Frederic Kenyon, *War Graves: How the Cemeteries Abroad will be Designed* (London: His Majesty's Stationery Office, 1918), 7.

8 Jay Winter, *Sites of Memory, Sites of Mourning: The Great War in European Cultural History* (Cambridge: Cambridge University Press, 1995), 24.

9 Daniel J. Sherman, "Art, Commerce and the Production of Memory in France after World War I," *Commemorations: The Politics of National Identity*, ed. John R. Gillis (Princeton: Princeton University Press, 1994), 188–9.

10 Philip Longworth, *The Unending Vigil: The History of the Commonwealth War Graves Commission* (Edinburgh and London: Constable, 2003), 33.

11 Bruce Scates, *Return to Gallipoli: Walking the Battlefields of the Great War*, (Cambridge University Press, Cambridge, 2006), 47. CWGC/2/2/1/68; CWGC/1/7/E/30.

12 CWGC WG 437; CWGC/1/1/9/C/11/.

13 In his report of January 1918 after a visit to the Gallipoli peninsula, the Director of the British Museum, Sir Frederic Kenyon, advised that "over the graves of all officers and men in the war cemeteries abroad [*there should be*] headstones of uniform dimensions, though with some variety of pattern" and every soldier's burial place would be recognized with "the rank, name, regiment and date of death of the man buried beneath it." *War Graves: How the Cemeteries Abroad will be Designed* (1918), 6–7; See also Laqueur, "Memory and Naming," 153.

14 CWGC DS/UK/ and CWGC/1/1/9/C/11/; CWGC DS/UK/ See for Burnet, "The War Cemeteries," 7–8.

15 CWGC DS/UK/; Burnet, "The War Cemeteries," 9.

16 CWGC WG 437; CWGC/1/1/9/C/11/.

17 CWGC WG 237 and CWGC WG 909; CWGC 1/1/9/A/5 and CWGC 1/1/9//C/15 (IWGC report dated 23 Feb 1927).

18 CWGC WG 437; CWGC 1/1/9/A/5; CWGC 1/1/9/C/15 (Letter of Kipling to Chettle 24/26 March 1926).

19 "Graves Where our Heroes are Buried," *The Queenslander*, 8 January 1916.

20 CWGC WG 437; CWGC/1/1/9/C/11/.

21 Ibid.

22 AWM 38 3DRL 8042-51.

23 Thomas J. Pemberton, *Gallipoli To-Day* (London: E. Benn, 1926).

24 AWM 38 3DRL 8042-51.

25 The work of this first group was then taken over in 1920 by the British Commonwealth War Graves Commission (CWGC), which quickly formulated plans for 31 cemeteries and five Allied commemorative monuments to be constructed on the peninsula.

26 The digitized version of C.E. Bean's Gallipoli Mission, 1st edition from 1948 is available through the AWM archives. See www.awm.gov.au/collection/C1416967. Charles Edwin Woodrow Bean, *Gallipoli Mission*; Bill Gammage, *The Broken Years: Australian Soldiers in the Great War* (Melbourne: Melbourne University Press, 2010, reprinted by Penguin Books (1981), 132.

27 The first of the archaeological surveys undertaken in 2003 by the joint Australian–Turkish team found that along the Second Ridge of the Anzac region there is a good possibility that some of the ANZAC dead are possibly buried in the Turkish trenches along this ridge as "unlike the Allies, the Turkish forces collected the unburied dead (after the evacuation) and placed them within nearby trench systems." David Cameron and Denise Donlon, "A Preliminary Archaeological Survey of the ANZAC Gallipoli Battlefields of 1915," *Australasian Historical Archaeology 23* (2005), 135.

28 C.E. Bean, *First World War Official Histories*, 1st ed. 1948, 72, 381–385. www.awm.gov.au/collection/C1416967) for digitized record at AWM Accession number RCDIG1069872

29 For the Australian government's request to obtain the area of the peninsula where the ANZAC troops first landed see TNA FO 608/116, 3 April 1919.

30 AWM38 3DRL 8042-51.

31 AWM38 3DRL 8042–51 (10 October 1919, "Report from Commonwealth of Australia to the Secretary of the Imperial Graves Commission.")
32 Ibid.
33 C.E. Bean, First World War Official Histories, 63.
34 AWM38 3DRL 8042–51.
35 Ibid. (27 July 1922 Letter to Mr. Knowles.)
36 Ibid.
37 Ibid., Gooding, *Gallipoli Revisited*, 101.
38 For the Treaty of Lausanne, see https://wwi.lib.byu.edu/index.php/Treaty_of_Lausanne.
39 Peter W. Stanley, *Quinn's Post: Anzac, Gallipoli* (Sydney: Allen & Unwin, 2005), 188.
40 B. West, and Curtin University of Technology, *Down the Road: Exploring Backpacker and Independent Travel*. Australia Research Institute (2005). Perth, WA: API Network, Curtin University of Technology, 28–9.
41 Jenny Macleod, "The Gallipoli Centenary: An International Perspective," *War Memory and Commemoration*, ed. B. West (London and New York: Routledge, 2017), 98–9.
42 Uluğ İğdemir, *Atatürk ve Anzaklar* (Ankara: Türk Tarih Kurumu, 1978).
43 Kahraman Şakul, ed., *Cihan Harbinde Gelibolu ve Mirası/Gallipoli in the Great War and its Heritage Exhibition Catalog* (İstanbul: İstanbul Şehir University, 2016), 10.
44 West, 28–9.
45 AWM38 3DRL 8042-51 (Letter from Hughes, Commander of the Australian Imperial Forces, 18 December 1918.
46 AWM38 3DRL 8042–51. Bean also noted the remade ANZAC graves. C.E. Bean First World War Official Histories, 63–4.
47 TNA W095/4954; WO 32/5640 Bruce Scates, *Return to Gallipoli: Walking the Battlefields of the Great War* (Cambridge: Cambridge University Press, 2006), 18.
48 AWM 38 3DRL 8042-51.
49 Ibid., (Report from Bean sent to AIF headquarters).
50 Ibid.
51 Ibid., (Letter from Hughes to Bean from Kilitbahir 7 Nov 1919).
52 Kenyon, *War Graves*, 10.
53 Ibid., 11.
54 Pemberton, 20.
55 David Dutton, "'Docile Supernumerary': A French Perspective on Gallipoli," in *Gallipoli: Making History*, ed. J. McLeod (London and New York: Frank Cass, 2004), 86.
56 Eleanor van Heyningen, "Helles: The French in Gallipoli," *Joint Imperial War Museum/Australian War Memorial Battlefield Study* (London: Imperial War Museum, 2000).
57 Dutton, 90.
58 For a discussion of the French involvement in fortifications along the Dardanelles and the Bosphorus see L. Thys-Şenocak and G. Dörter, "Fortifications in Turkey: The Ottoman Frontier Fortresses of Seddülbahir and Upper Rumeli Kavak," in *Conservation of Cultural Heritage in Turkey*, ed. Z. Ahunbay, et al. English ed. L. Thys-Şenocak and E. Yıldırım (İstanbul: Ege Yayınları/ICOMOS Turkey, 2016), 365–86.

59 Philip Longworth, *The Unending Vigil: A History of the Commonwealth War Graves Commission, 1917–1984* (S. Yorkshire: Leo Cooper 2003), 11.

60 For the tradition of French burials in ossuaries see Annette Becker, "From Death to Memory: The National Ossuaries in France after the Great War," *History and Memory*, vol. 5, no. 2 (Fall–Winter, 1993), 32–49, D.M. Hughes, "The French Army at Gallipoli," *The RUSI Journal, 150* no. 3 (2005), 64–7.

61 Tonie Holt and Valmai Holt, *Major & Mrs Holt's Battlefield Guide to Gallipoli* (S. Yorkshire: Leo Cooper, 2000), 118–19. The passage is from Victor Hugo's song "Hymne":
Gloire à notre France éternelle,
Gloire à ceux qui sont morts pour elle,
Aux martyrs, aux vaillants, aux forts,
A ceux qu'enflamme leur exemple,
Qui veulent place dans le temple
Et qui mourront comme ils sont morts.

62 CWGC WGC/1/1/9/C/11 (E.E. Hughes, IWGC Deputy Director of Works, to C.E. Bean, 22 March 1919, regarding progress with the cemeteries of the peninsula).

63 CWGC/1/1/9/C/12.

64 Pemberton, 43–4.

65 J. McQuilton, "Gallipoli as Contested Commemorative Space," in *Gallipoli: Making History*, ed. J. Macleod (London and New York: Frank Cass, 2004), 150–8; See also J. Winter, *Sites of Memory, Sites of Mourning*.

66 McQuilton, 151.

67 TNA FO 608/116.

68 Bilal N. Şimşir, *İngiliz Belgelerinde Atatürk (1919–1938)* vol. 4 (Ankara: Türk Tarih Kurumu Yayınları, 1984), 464.

69 McQuilton, 151.

70 Pheroz Unwalla, "Re-imagining Gallipoli: Imperial Pasts and Foreign Presence in a History of Turkish National Remembrance, 1923–2013" Ph.D. diss., (SOAS, 2014), 31.

71 Richard Reid et al., "Remembering Gallipoli, *Anzac Battlefield*, 221.

72 For an analysis of the different ways in which Gallipoli is remembered in Britain and Ireland see Jenny Macleod, "Britain and Ireland: Gallipoli Day or Anzac Day," in *Gallipoli* (Oxford: Oxford University Press, 2015), 126–53. Macleod notes on page 151 that the first official recognition of the Republic of Ireland of the Irish soldiers who fought with the British in Gallipoli was in 2010 when the President of Ireland Mary McAleese visited Gallipoli for the unveiling of a memorial plaque at Green Hill Cemetery.

73 Ibid.

74 Ibid., 129.

75 Holt and Holt, 95.

76 R. Fuchs, "Sites of Memory in the Holy Land," *Journal of Historical Geography* 30 (2004), 657.

77 Holt and Holt, 99.

78 For an in-depth study of Burnet see Ahenk Yılmaz, "Memorialization on War-Broken Ground: Gallipoli War Cemeteries and Memorials Designed by Sir John James Burnet," *Journal of the Society of Architectural Historians* 73, no. 3 (2014), 328–46.

79 See also A. Service, *Edwardian Architecture and Its Origins* (London: Architectural Press, 1975), 192–215; A.S. Gray, *Edwardian Architecture* (University of Iowa Press: Iowa City, 1986), 128–32; D. Walker, "Monumental Modernist," *RIBA Journal*, 988 (1991), 20–5.
80 AWM 38 3DRL 8042-51 (Hughes to Bean, 13 Dec 1919).
81 CWGC 1/1/9/C/14 and CWGC 1/1/12/40 (Letter to Kenyon 15 October 1919).
82 CWGC 1/1/9/C/14 and CWGC 1/1/9/11 and DS/UK/8 (John J. Burnet report to IWGC 30 June 1919).
83 AWM 38 3DRL 8042-51 (Hughes to Bean, 13 December 1919).
84 Ibid.
85 Ibid.
86 CWGC 1/1/9/C/14 and CWGC 1/1/12/40 and CWGC 1/1/7/E/20 (Letter from John Burnet 30 June 1919).
87 Ibid.
88 Ibid.
89 Catherine Moriarty, "Christian Iconography and the First World War Memorials," *Imperial War Museum Review* 6 (1992), 74.
90 Fuchs, 651.
91 This assumption has proven to be correct; the gardener of the French Cemetery at Morto Bay reported that in the spring of 2014 vandals had destroyed some of the decorative ceramic reliefs in the monument which had Christian symbols.
92 Winter, 97.
93 Yılmaz, 330; see also Patricia Rubertone, "Monuments and Sexual Politics in New England Indian Country," in *The Archaeology of Colonialism: Intimate Encounters and Sexual Effects*, ed. Barbara Voss and Eleanor Conlin Casella (Cambridge: Cambridge University Press, 2012).
94 CWGC 1/1/9/C/14 and CWGC 1/1/9/11 and DS/UK/8. (Burnet to IWGC, 30 June 1919, Burnet Report Catalogue 408, 8–9.).
95 CWGC 437/4 ("The Monument on Cape Helles: A Landmark for All Time," *The Builder*, 9 May 1924).
96 CWGC 21361 H58 003 for the description of the foundations of the Cape Helles monument and WG 1687/4 Catalog 550 Box 1103 for the mention of a "raft" being prepared for the Cape Helles monument from the Decauville rail ties.
97 At the CWGC archives I was not given permission to copy the original drawing No. CH7A, Item 44 which shows the placement of the repurposed Decauville rails. Notes on the drawing indicate that the CWGC's copy is actually a traced drawing of the original which was made by H.L. Yeo on 31 October 1924. It shows clearly the placement and length of the different rails and where they were used for reinforcement of the monument's base.
98 Ibid.
99 Ibid.
100 CWGC 1/1/9/C/14 and CWGC 1/1/9/11 and DS/UK/8 (Letter from John Burnet 30 June 1919).
101 CWGC 437/1 and CWGC 1/1/7/E/20 (E.E. Hughes Deputy Director of Works IWGC to Imperial War Commission, 22 Feb 1922, "Marking points of interest Gallipoli.") Ilgardere is also referred to as UlgenTepe in these reports.
102 Pemberton, 46.
103 CWGC WG1617 (Burnet report of 30 June 1919).

104 Ibid., 4–5.
105 Ibid.
106 Ibid.
107 CWGC WG 1617; CWGC ADD 1/1/12 ("Report as to the Best Position for a British National War Memorial at the Dardanelles" presented by Lieut-General Sir Aylmer Hunter-Weston to the Secretary of the National Battlefields Memorial Committee in the House of Commons, 14 June 1921).
108 This is a reasonable and prescient concern, considering the 2015 theft of the bronze plaques at ANZAC Cove.
109 CWGC ADD 1/1/12 ("Report as to the Best Position for a British National War Memorial at the Dardanelles"), 4.
110 Ibid., 5.
111 CWGC 437/4 ("The Monument on Cape Helles: A Landmark for All Time," *The Builder*, 9 May 1924).
112 Fuchs, 651.
113 Winter, 96.
114 C.E. Bean, ed., *The Anzac Book: Written and Collected in Gallipoli by the Men of Anzac* (London: Cassell, 1916), 21.
115 David W. Lloyd, *Battlefield Tourism: Pilgrimage and the Commemoration of the Great War in Britain, Australia and Canada, 1919–1939* (Oxford and New York: Berg, 1998).
116 Chris Ryan, ed., *Battlefield Tourism: History, Place and Interpretations* (Oxford: Elsevier, 2007), 5. In his explanation of the evolution of hot to cold war tourism sites, Ryan refers to the article in that volume by Mark Piekarz, "It's Just a Bloody Field! Approaches, Opportunities and Dilemmas of Interpreting English Battlefields," 29–48, and notes that "such sites become less hot as hostilities cease, time passes, and eventually the direct combatants die and pass away. The events then pass into history and subsequent generations interpret them in terms of their own needs. Such interpretations can range between the poly-vocal and the uni-vocal as those subsequent generations seek to either confirm their own status and homogeneity by perceiving others through a gaze of 'otherness' or alternatively embrace more inclusive interpretations."
117 Lloyd, 4.
118 H.J. Greenwall, *Daily Express*, (23 Sept 1919).
119 *Traveller's Gazette*, LXX, (8 March 1920), 3.
120 CWGC G/928/G/V CWGC 1/1/16/25 (14 July 1928, Confidential communication about Gallipoli pilgrimage).
121 Ibid.
122 Ibid.
123 CWGC ACON 149: CWGC 1/1/16/25 "St Barnabas Pilgrimage to Salonika and Gallipoli 1928." Seven organized pilgrimages were made between 1920 and 1930; in 1925 the SS *Otranto* landed passengers at Cape Helles. In 1926 and 1928, two pilgrimages were organized by St Barnabas.
124 CWGC 270/13/28.
125 CWGC 1/1/7/E/35.
126 For records of the French Graves in Gallipoli see CWGC WG 1188/6; CWGC/1/1/8/15 (Letter to the Deputy Director of Works, IWGC (Gallipoli) written from Mudros 16 March 1922).

127 Ibid.; CWGC 15056/12004/44.
128 CWGC 437/7; CWGC/1/1/10/G/1 (Letter dated 26 June 1931).
129 CWGC WG 437/7; CWGC/1/1/10/G/1.
130 Ibid.
131 TNA CM 13/28 (Letter dated 23 July 1970 and GKG 318/4/3 for confidential consular letter of 7 July 1970).
132 Unwalla, 30–1.

5 Commemoration begins for the Ottoman martyrs

> This task has dragged on. It is our right to see a magnificent statue of the victorious Mehmetçik along with the monuments erected by the defeated nations at Çanakkale. Everywhere in Çanakkale there should be thoroughly Turkish monuments. . . . However we want to see the heroic deeds of great Mehmet shown in an appropriate manner. When the foreigners, who visit their own monuments, want to place bunches of flowers at a Turkish monument out of respect, they are surprised when they cannot find one. This is very sad. It is our clear duty to construct a grave, a monument for the Mehmetçik.[1]
>
> Kazım Özalp Pasha, Speaker of the Turkish Grand National Assembly, August 1933

By 1931 plans were underway to reconstruct the Mehmet Çavuş Monument and Cemetery at Anafartalar on the Gallipoli peninsula, and a year later the architect Kömürcüzade Asım had finished the drawings for a monument which would honor the victorious Mehmetçik, an answer to Kazım Özalp Pasha's request, above.[2] The monument had originally been built in 1915 by Commander Mehmet Şefik Aker at the site of Cesarettepe to honor sixty of the Ottoman soldiers who had been killed there in his 19th Division.[3] Initially a quite rudimentary pyramid of concrete surmounted by artillery shells, the monument would be clad in marble by 1934; new inscriptions were also placed on its façades to commemorate the different dates of invasions, bombardments, and evacuations of the Allied forces from the Gallipoli peninsula. The Mehmet Çavuş Monument, in its refurbished state, would serve as the site for all official ceremonies conducted by the Turks on the peninsula for the next thirty years (Figure 5.1).[4]

The less organized commemoration of Ottoman troops who died or were wounded at Gallipoli, when compared to the more systematic efforts undertaken by the Allied countries to bury and commemorate their Gallipoli dead immediately after the war ended, has been noted in many studies of the campaign.[5] Some authors have concluded that a comparative lack of presence of Turkish commemorative monuments on the peninsula, particularly in the years immediately after the war, is an indication that the Ottomans were uninterested in remembering the soldiers who had died during these battles. Alan Moorehead, the author of one of the more popular histories of the Gallipoli campaign, claimed in 1956 that Necmettin

Halil Onan's poem "Dur Yolcu/Stop Wayfarer", inscribed above the straits, was the only monument to the Turkish dead. The discrepancy between the numbers of Allied monuments and those of the Turks, he explained, was because "The Turks find this preoccupation with the dead strange since their own soldiers who died at Gallipoli were buried in anonymous communal graves, and until recently almost their only memorial was a legend picked out by large white letters on the hilltop above the Chanak."[6] The British reporter John Laffin concluded, in his 1974 report for the Royal British Legion journal, that "The Turks have no cemeteries: their 66,000 dead were buried where they fell or in mass unmarked graves."[7]

Contemporary observers such as Lord Stoppard, who reported on the 1928 St. Barnabas Pilgrimage to Salonika and Gallipoli (discussed in Chapter 4) had noted while he was traveling on the penininsula that "material is being collected at Helles for a Turkish memorial but it has not yet been decided where to erect it. While going around I saw two small Turkish monuments built of the poorest materials imaginable."[8] Thomas James Pemberton, who was working on the peninsula in 1926 with the CWGC, remarked upon the differences between the commemorative approaches of the Turks and the Allied forces, "The British seem to have carried this idea of permanent sepulture even further than other modern nations. The Turks, whose dead lie in Gallipoli soil in equal numbers to those of their erstwhile enemies, have made no attempt whatever to mark their places of burial."[9] Later, in his book *Gallipoli To-day*, Pemberton did acknowledge that the Turks had raised a "rough monument of victory," but that "nothing could have

Figure 5.1 The reconstructed Mehmet Çavuş Monument at Arıburnu on the Gallipoli peninsula.

Credit: Archive of L. Thys-Şenocak

been more modest than this pyramidal block of concrete, surmounted by five live shells, and standing only some fifteen feet in height." His overall assessment of the Turkish commemorative approach on the peninsula was that "the Turks raised no other memorial to their dead. Here and there are to be found Turkish cemeteries, but they are not marked nor intended to be remembered."[10]

The monument Pemberton was likely describing was located at the Nek, one of the most contested sections of the Gallipoli battlefields, where on 7 August 1915 the Australians were stopped from advancing. While C.E. Bean had noted in 1919 a white obelisk here erected by the Ottomans and mentioned that "the Turks were very proud of their achievement", Pemberton was not impressed and claimed it bore no resemblance to the white marble monuments that were being erected by the CWGC in the years just after the war. "Its surface is rough and it is partly formed of snipers' masks—a strange contrast to the costly and beautiful monuments of the vanquished."[11] Later the Turks would redesign the monument, which today retains only its original base; after the war, the sniper masks that Pemberton found unfitting for a memorial were replaced with the more traditional marble (Figure 5.2).[12]

Given these observations made by both Turkish and foreign observers in the years just after the war ended, and for several decades thereafter, questions arise. Did the Ottomans, and then the Turks of the new republic, make less of an effort to commemorate their dead and missing in the two decades after the war, when compared to the Allied countries? Did they somehow care less than Europeans about remembering and mourning their fallen? Were they simply lacking an architectural vocabulary to use for wartime memorials and commemorative architecture? To answer these questions we must move beyond the actual

Figure 5.2 CWGC (lower) and Turkish (upper) commemorative monuments at the Nek, where on 7 August 1915 the Australians were stopped from advancing.

Credit: Courtesy of the CWGC archives

landscape of the peninsula and its surroundings, and the decades before and just after World War One, to examine several cultural and political factors which shaped the architectural responses of the Ottomans (and then Turks) to the commemoration of their dead. Both the burial traditions and commemorative practices of the late Ottoman era, the architectural styles and resources that existed in the Ottoman Empire during the years immediately preceding World War One, provide some insight and help to contextualize the earliest commemorative structures built on the peninsula.

British and French forces returned to the Gallipoli peninsula by the fall of 1918, just as the Ottomans were becoming embroiled in another conflict: the War of Independence. The Ottomans' limited access to, and uncertain ownership of, the peninsula during the first years after the war ended are important factors to consider. Some authors such as Sibel Baykut have concluded that:

> Neither the people nor the government in Turkey had the opportunity to go back and think about commemorating Gallipoli because it was only one of six fronts that Turkey had to fight in those years, each of which had its own tragedies and deserved mourning. Moreover, the people in Turkey had another three years of fighting after WWI for their independence. Thus, Gallipoli would simply have to wait.[13]

While Gallipoli did wait almost a half-century for the construction of a major commemorative monument on the peninsula—the Çanakkale Martyrs Memorial—the efforts made by the Ottomans during and immediately after the war, on and off the Gallipoli peninsula to commemorate the dead through smaller monuments and plans for larger ones should be recognized. Families in the Ottoman Empire who lost their loved ones in the Balkan Wars, at Gallipoli, and in other battles during World War One and the War of Independence certainly grieved for their dead; parents, comrades, and commanders sought ways to make sense of the violence, death, and turmoil that accompanied the upheavals and ruptures in daily life as the empire came to an end. The failure of rows of Ottoman headstones and monumental white obelisks to appear immediately after the battles on the peninsula ceased should not lead to the conclusion that the Ottoman dead and missing were forgotten by the state and "not intended to be remembered."[14]

Atatürk himself, during a visit to the peninsula in 1935, noted that there was no Turkish commemorative monument of comparable size to that of the former Allied forces but reasoned that there was time for this:

> True, we too shall build monuments to commemorate our martyrs, but this is a matter of time and opportunity. Still, believe me that our soldiers have built the greatest memorial of all by keeping this land within the borders of Turkey.[15]

To begin to answer these questions, it is helpful to examine the commemorative practices of the Ottomans during the final decades of the empire, and how concepts such as martyrdom had evolved and were expressed in the empire's architecture just before World War One. In *Sites of Memory, Sites of Mourning*, Jay Winter writes about the many different purposes that commemorative architecture can have for the individuals, communities, and governments that fund and build these memorials. In Europe, after World War One began, commemoration and the construction of memorials for commemorative purposes constituted an act of citizenship and also created spaces where the dynamics of power between the state and the community could play out. The European war memorials of World War One functioned as "sites of symbolic exchange, where the living admit a degree of indebtedness to the fallen which can never be fully discharged." As such, they provided spaces to grieve collectively, to "help individual people—mothers, fathers, wives, sons, daughters, and comrades-in-arms—to accept the brutal facts of death in war."[16]

In the late Ottoman Empire, prior to World War One, there were very few state-commissioned public monuments to commemorate collectively those who died fighting for the sultan. During the reign of the Ottoman Sultan Abdülhamid II (r. 1876–1909), the practices of state-organized burials and collective commemoration were just beginning. By the late nineteenth century the use of the term "martyr" (*şehit*) on individual Ottoman tombstones was becoming more frequent, but it was not used exclusively for those who had died fighting for the sultanate or the Islamic faith; it could also be used for a woman who had died in childbirth, or someone who had passed away due to an illness or natural disaster.[17] With the fatalities of the Greco-Ottoman War of 1897, and then the Balkan Wars of 1912–13, the Ottoman state began to utilize the term *şehit* in a more defined and consistent manner to designate those who had died in battle, fighting for the empire. As Eldem has pointed out, at this time in the Ottoman Empire "The emergence of an official (i.e. state-defined) notion of martyrdom was the result of the rise of the modern state, and more particularly, the development of state nationalism."[18]

Pre-WWI Ottoman memorials: Istanbul

To understand the historical and architectural context in which the commemorative monuments built by the Ottomans on the Gallipoli peninsula were created, it is useful to examine some examples of memorials—their stylistic choices and iconographic meanings—that were constructed just prior to World War One in Istanbul and Edirne, as these memorials would shape the Ottoman choices for commemorative monuments on the Gallipoli peninsula both during and after the war there. Two examples from the late Ottoman Empire—the *Abide-i Hürriyet* (Freedom Monument) and the *Tayyare Şehitleri Abidesi* (Aviation Martyrs Monument), both in Istanbul—illustrate that these public structures were intended to impart a different message from their equivalents in Europe. A large percentage of the late nineteenth-century Ottoman commemorative memorials in Turkey that are still standing have been restored in recent decades, so it is often difficult to determine the original design of many of these. However, for the public monuments or memorials for which the original design is known, variations of the obelisk, or "column monuments", were the most common stylistic choices. Occasionally, more eclectic Orientalizing styles typical of late nineteenth-century architecture were employed, as with several memorials erected for the Ottoman dead in Edirne shortly after the Balkan Wars.[19] Some of the more traditional European designs for commemorative monuments did not become popular in the Ottoman Empire as they evoked what was deemed as inappropriate symbolism. As the late nineteenth-century historian Ottoman Ahmed Lütfi Efendi pointed out, the triumphal arch, frequently used for commemorative structures in Europe, was a building typology perceived as being too European. By contrast, the obelisk was more acceptable and familiar, and resonated with the *"nişan taşları,"* a stone that served as the target the Ottomans had used for centuries for archery practice.[20]

Stylistic and iconographic choices made by most architects of the earliest commemorative Ottoman monuments generally followed the architectural trends for commemorative monuments in Europe, which was itself quite traditional in the 1920s. As Borg has noted, this type of architecture remained conservative and rooted in traditional forms as it "undoubtedly reflects the official patronage which sponsored the building of the memorials, for such patronage is normally conservative in its choice of styles."[21] The *Abide-i Hürriyet* or Freedom Monument, erected in Şişli, a central district of Istanbul, was the first collective public memorial built by the Ottoman state to commemorate those who had died in battle. It was dedicated to the officers of the "Freedom Army": men who had lost their lives during a counter-revolution in 1909 while defending the government of the Young Turks. According to the burial records, these men were officially considered and named by the state as martyrs.[22] The memorial was designed by Mimar Muzaffer, who had been the city architect of Konya in central Anatolia. Muzaffer had competed for the commission with many well-known architects working in the late Ottoman Empire, including Kemalettin Bey, Vedat Tek, Konstantin Kiryaki, and Alexandre Vallaury (Figure 5.3).[23]

Figure 5.3 The *Abide-i Hürriyet* or Freedom Monument, designed by Mimar Muzaffer, was erected in Şişli, a central district of Istanbul.

Credit: Archive of L. Thys-Şenocak

With its Ottoman flags, sultan's *tuğra* (insignia), the barrel of a cannon used in lieu of an obelisk, and the unusual placement of an underground mosque and *mihrab* (a prayer niche indicating the direction of Mecca), the Freedom Monument stands out as an interesting mixture of monumental commemorative architecture which drew from typologies in Europe but which also incorporated several elements of the late nineteenth century/early twentieth century Ottoman revivalist architecture that was state-supported and popular in the final years of the empire. At the opening ceremony for the *Abide-i Hürriyet*, Enver Bey of the Young Turk government proclaimed that "Muslims and Christians lay side by side, a token that they, living or dead, were henceforward fellow patriots who would know no distinction of race or creed."[24]

The *Abide-i Hürriyet* does commemorate collective sacrifice and it therefore shares this feature of memorialization with World War One monuments built throughout Europe, but its main purpose was not to foster a feeling of citizenship among those who came to mourn the dead as was the case with so many European commemorative structures. Rather, the *Abide* was built to cultivate the allegiance of former Ottoman subjects to the new government of the Young Turks. More than any other purpose, this memorial served as a statement of power and victory while it commemorated a particular event in the tumultuous political landscape of a transitional government.[25]

Another larger Ottoman monument to the fallen built in Istanbul is the *Tayyare Şehitleri Abidesi* (Aviation Martyrs Monument) located in the Fatih district of Istanbul (Figure 5.4). The construction of this memorial began in April 1914 and it was designed by one of the masters of Ottoman Revivalist architecture, Vedat Tek (1873–1942). Enver Pasha, the Minister of War at the time, laid the foundation stone for the memorial in the spring just before the major battles of the Gallipoli peninsula began. The monument was built to honor three Ottoman pilots who had lost their lives while flying from Istanbul to Cairo and Alexandria on a 2,500 km reconnaissance mission across Ottoman territory. A tribute to the first Ottoman aviators, the single white marble column at the center of the monument stretches to a height of 7.5 m and is flanked by bronze medallions. For the central section of the memorial, the architect used a broken column to symbolize the journey of the pilots and their lives which were truncated by the accidents. Ultimately, like its predecessor, the *Abide-i Hürriyet*, Vedat Tek's *Tayyare Şehitleri Abidesi* is a monument that commemorates collectively, but those remembered are a select group of distinguished individuals who represented the accomplishments of the late Ottoman government. In addition to the creation of the memorial, the special commemoration date of 27 January was established in 1925 to honor those who lost their lives in aviation accidents.[26]

While the men commemorated by the Aviation Martyrs Memorial and the Freedom Memorial were all officers, not common soldiers, these late Ottoman commemorative projects in Istanbul mark a turning point in the Ottoman government's approach to the ways in which its soldiers and civil servants were commemorated. Eldem notes, "The idea of replacing individual gravestones of

Figure 5.4 The *Tayyare Şehitleri Abidesi* or Aviation Martyrs Monument located in the Fatih district of Istanbul was designed in 1914 by Vedat Tek (1873–1942).

Credit: Archive of L. Thys-Şenocak

war heroes with a collective monument was clearly the first instance of the application of the model of western war commemorative monuments to an Ottoman context. The national symbolism of the Freedom Monument constitutes one of the first concrete examples of the transition from the equation of 'nation = state' to that of 'nation = people'."[27] There would, however, continue to be diverse approaches taken by the Ottoman state, and bereaved families to the design and intended meanings of commemorative architecture, particularly outside of Istanbul.

Memorials of the Balkan Wars: Edirne

Only two years prior to the commencement of the battles on the Gallipoli peninsula, the Ottomans had buried their dead from the Balkan Wars (1912–1913) and had made plans to commemorate these soldiers. In the Thracian city of Edirne, formerly known as Adrianople, many memorials and cemeteries for the dead of the Balkan Wars were established, employing a wide range of styles and symbolism. Different meanings were attributed to individual and collective memorials. While there are several cases of collective burials with only the name of the regiment leader appearing on the tombstone/epitaph, there were also collective memorials constructed by surviving members of a regiment who wished to honor all their fallen comrades.[28] Some of these, such as the *Beş Şehitlik Anıtı* (Five Martyrs Memorial) and the *Söğütlük Jandarma Şehitliği* (Söğütlük Gendarmerie Memorial) erected in March 1913, listed all the names of the dead on the monument, regardless of rank.[29]

The deaths of individual soldiers were also remembered with monuments built by family members. Departing from the more reconciliatory tone that was common among the commemorative monuments in Europe at this time, the *İntikam Anıtı* (Memorial of Revenge) is one interesting example of a memorial from this era which had a much less forgiving message. Erected in the garden of the Selimiye Mosque in Edirne in 1914 by Fuat Pasha, who had lost his son Reşit Bey during the Bulgarian invasion of the city, the monument marked the day and bitter memory of the brutal death and mutilation of his son by his attackers on 9 July 1913.

The diversity of styles in Ottoman commemorative architecture prior to World War One is particularly evident in the Balkan Wars cemeteries of Edirne. These provide important evidence that, just prior to the outbreak of war in 1914, both the Ottoman state and the families of those who had died in battle made efforts, during and immediately after the fighting had ceased, to remember their dead by constructing many different types of memorials. While a significant amount of restoration was done on the Edirne memorials in 1935, and again in the mid-1960s, archival photographs and the remains of some of the original commemorative structures indicate that the latter were often designed as small obelisks, frequently decorated with cannon balls or artillery shells. As with the Istanbul commemorative monuments built just before World War One, European designs for Ottoman memorials, and occasionally European architects themselves, provided the stylistic inspiration for the Ottoman monuments built during and immediately after the Balkan Wars.

The project for the Söğütlük memorial was jointly prepared by the Ottoman architect Talat Bey, the governor of Edirne Hacı Adil Bey, and a French engineer named Dupré.[30] With its architectural references to Egyptian pyramids and Chinese pagodas, the Söğütlük memorial incorporated some of the elements of funerary architecture popular in France at the time. Its orientalizing design, however, was not appreciated by some critics, such as Rıfat Osman, who wrote in 1927 that a fountain with all the names of the dead inscribed on it would have been more appropriate and meaningful.[31]

Figure 5.5 The Italian War Monument was erected in 1913 on the upper level of the fortress of Seddülbahir at Cape Helles to commemorate the Ottoman victory over the Italians in the Tripoli War (1911–12).

Credit: Archive of L. Thys-Şenocak

On the Gallipoli peninsula itself, the Italian War Monument was erected in 1913 (the year just prior to the commencement of Allied–Ottoman conflict in the Dardanelles) in the upper level of the fortress of Seddülbahir at Cape Helles to commemorate the Ottoman victory over the Italians in the Tripoli War (1911–12). This modest victory monument, a thin column on which an artillery shell was placed, displayed just below the base of the shell the names of the soldiers who had fought against the Italian bombardments of the fortress during that war. It was still standing after the fortress was attacked a few years later in World War One (Figure 5.5).

Commemorating the Ottoman dead on the Gallipoli peninsula

While the battles on the Gallipoli peninsula were won by the Ottomans, the empire came to an end with the larger losses incurred in World War One. With the Armistice of Mudros, signed between the Ottoman Empire and the Allied forces on 30 October 1918, the control of the Dardanelles and the right to occupy the fortresses along the Straits, including the two fortresses of Kilitbahir and Seddülbahir on the Gallipoli peninsula, was handed to the former Allied countries. Part of the plan to divide the remains of the Ottoman Empire was outlined at the Paris Peace Conference in January 1919; at that time much of western Anatolia, including eastern Thrace and the Aegean islands of Gökçeada (Imbros) and Bozcaada (Tenedos) just opposite the Gallipoli peninsula, was intended to be given to Greece to support their *Megali* project, creating a new Hellenic Empire composed of many of the Ottoman lands that were once part of the Classical world.

It was only with the Turkish War of Independence, which began on 19 May 1919 and ended more than four years later, that Allied plans for the dismemberment of the Ottoman Empire were finally thwarted. With the end of the War of Independence on 24 July 1923, and the establishment of the Turkish Republic three months later on 29 October 1923, relative stability and a clearer idea of the nature of sovereignty over territories that had remained from the Ottoman Empire emerged. However, several aspects of the newly proclaimed Turkish nation were in flux, including the definition of citizenship and the role that religion and the state would play in commemorative practices.

For the Allied countries, World War One ended on 11 November 1918, and many of the commemorative projects (including those on the Gallipoli peninsula) began, or their progress accelerated, as the result of a more peaceful atmosphere at home on the European mainland. But for the Ottomans, the peace gained after the evacuation of the Allied forces from the Gallipoli peninsula was very brief. As another conflict, the War of Independence, began only five months later, the upheaval and instability that characterized the final years of the late Ottoman government, coupled with the uncertainty over who would actually end up owning the Gallipoli peninsula, made it difficult for the Ottomans to organize any kind of major building campaign to erect commemorative structures on a scale comparable to the Europeans.

Experience had also shown the Ottomans that some of the Balkan Wars memorials, such as the Kartaltepe Martyrs Memorial, constructed in November of 1912, and the Kurtkale Police Memorial, built for thirty of the soldiers who had died in the early days of the wars, ended up within the redefined Greek and Bulgarian borders respectively. This unforeseen development would have created apprehension among the Ottomans about deciding where to erect major memorials for the Gallipoli dead immediately after the war, since the ultimate ownership of the peninsula was still undecided and a number of countries had an active presence there during the 1920s. In addition to the Imperial War Graves Commission and the large group of laborers they had recruited for their building projects, French archaeologists had returned to excavate at Cape Helles, along with garrisons of French Senegalese soldiers stationed at several locations around the peninsula including the Seddülbahir and Kilitbahir fortress. Several Greeks had also returned during the early part of the decade to move into what remained of the houses in the villages.

As Mithat Atabay has noted, some early monuments to the Ottoman dead had been constructed on the peninsula during the time of the battles or immediately after the evacuation of the Allied forces in 1916.[32] These included the Maltepe Third Army Corps Telegram Squadron Monument and Cemetery,[33] the Çataldere Monument, the Last Arrow Monument, the 16th Division Monument, the Mehmet Çavuş Monument, the Kireçtepe Memorial, and the Anafartalar Victory Monument (Figure 5.6).[34] These early memorials, at least the first versions, were generally built quite hastily by the troops themselves from local building materials and frequently included ordnance such as artillery shells and other weaponry as decoration. The 16th Division Monument, which commemorated the Ottoman martyrs who took part in the Kanlısırt battle of August 1915, was somewhat more elaborate. C.E Bean recalled seeing several of these Ottoman monuments (Figure 5.7):

> I saw now, with something of a shock, standing out near the site of the vanished tree, a white obelisk—a monument (Hughes told us) put up by the Turks to mark the spot at which they had stopped the terrific August thrust. Away on the ridges nearly a mile beyond it, at the Nek where we had also been stopped, we could see another monument, and after we noticed a third, near North Beach.[35]

There are few traces of the commemorative monuments that were built on the peninsula by the Ottomans to honor their dead, as these were destroyed by the Australians, British, and French when they returned to the peninsula to embark on their own commemorative and archaeological projects in the immediate post-war years. The Anafartalar Victory Monument erected in 1916 to commemorate the 20 December 1915 Allied evacuation of this part of the peninsula was photographed

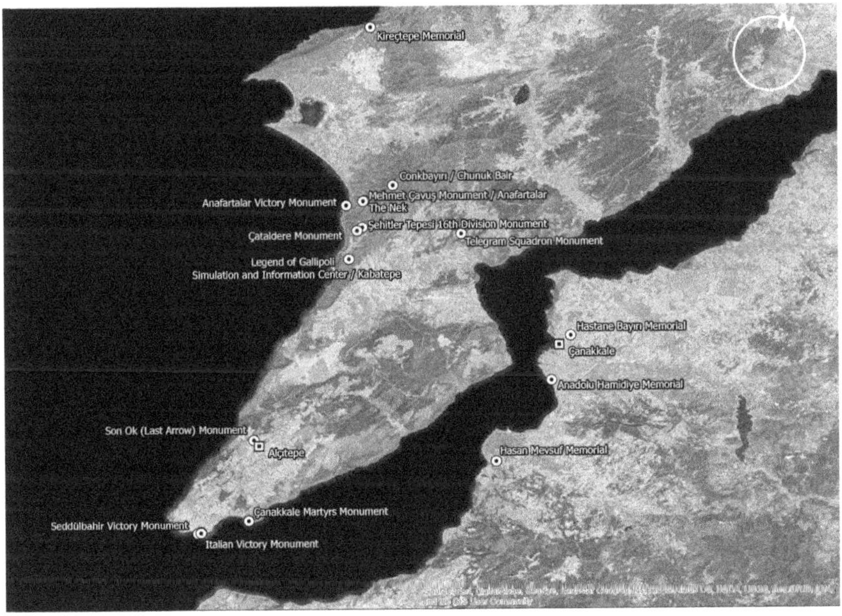

Figure 5.6 Map showing the locations of several early commemorative monuments that were built by the Ottomans during or immediately after World War One.

Credit: Google Earth, Elifgül Doğan

Figure 5.7 One of the early Turkish victory monuments at Şehitler Tepesi, the 16th Division Monument, was erected shortly after the evacuation of the Allied forces.

Credit: AWM G01752

by C.E. Bean in 1919 and then blown up in November of that year by his team as part of the plan for the Allied forces to reorganize the cemeteries and commemorative monuments undertaken by Bean and his team. Bean also noted that, "The Turks had not buried their men as well as we did, but now I noted three Turkish cemeteries at Old Anzac – two on the southern side of Wire Gully near its opening towards Legge Valley,one of these well-kept, the other a single disheveled trench. A third lay in Legge Valley itself, and there were a few scattered tombs."[36]

The Seddülbahir Victory Monument had been constructed by the Ottomans in the Cape Helles region of the peninsula, near the Ottoman fortress, to commemorate the evacuation of the Allied troops from that section of the peninsula in January

Figure 5.8 The Seddülbahir Victory Monument, destroyed by the French in 1920, was built by the Ottomans near the fortress of Seddülbahir to commemorate the evacuation of the Allied troops from that section of the peninsula in January 1916.

Credit: Gunter Hartnagel archive

1916. This monument was destroyed by the French in 1920 before they recommenced their archaeological excavations in the region from 1921 to 1923 (Figure 5.8).[37] Another Ottoman memorial, erected on the peninsula in 1915 at Lone Pine, was also allegedly destroyed by Australian troops in 1918.[38]

C.E. Bean, who in 1919 was working to document the Australian graves on the peninsula, mentioned in his reports that the Ottoman Mapping Directorate had conducted a documentation campaign of the peninsula from December 1915 through January 1916 under the leadership of Brigadier General Mehmet Şevki Pasha.[39] The result of Mehmet Şevki Pasha's work was a set of 43 very detailed maps of different parts of the peninsula, produced on a scale of 1:5000. These maps provided a wealth of information about the existence and location of many features on the peninsula, including the Ottoman and Allied artillery, water supplies, and fortifications. It also showed 28 cemeteries, created for Ottoman soldiers who had died during the battles fought there.[40] Many of these cemeteries were mass burials, but there were some individual graves indicated on the maps as well, along with mining tunnels, pits, camps, and even a bakery.[41] A century later, the information from the Şevki Pasha maps, along with the research about the names, ranks, birthplaces and place of martyrdom for 50,687 Çanakkale missing and dead on the Ottoman side was compiled by the Turkish Ministry of Defense and the Turkish Armed Forces General Staff into a database and used from 2005 to 2014 by the Turkish Ministry of Forestry and Water Affairs to commemorate many of the Ottoman martyrs who had not been recognized previously and as a catalyst for constructing several actual as well as symbolic cemeteries and memorials on the peninsula.[42]

Commemorating the Ottoman dead off the Gallipoli peninsula

In addition to the memorials and cemeteries created by the Ottomans on the Gallipoli peninsula during the war, it is important to note that there were several cemeteries and commemorative structures built for the Gallipoli martyrs in locations off the peninsula. Ottoman archival records attest to at least three cemeteries established in 1916 on the Asian side of the Dardanelles, all in or near Çanakkale: the Hastane Bayırı Memorial, the Anadolu Hamidiye Memorial, and the Hasan Mevsuf Memorial.[43] One particularly significant example of a large Ottoman cemetery and commemorative monument built during the years of the Gallipoli battles for soldiers who were wounded or died on the peninsula was located in the town of Biga, 100 kilometers to the north of Çanakkale on the Asian shore of the Marmara Sea. The Namazgah military cemetery in Biga was the burial site for over 200 Ottoman soldiers who died in the earliest battles on the Gallipoli peninsula or during treatment in the military hospital adjacent to the cemetery (Figure 5.9). The Biga hospital was where many of the wounded and dead of the battles at Seddülbahir and Anafartalar were taken to recover in 1915. According to the Ottoman officer in charge of the fortifications in Çanakkale, the Ottoman commemorative structure and cemetery at Biga were specifically built "to remind the future generations of the bravery and the efforts of 236 dead

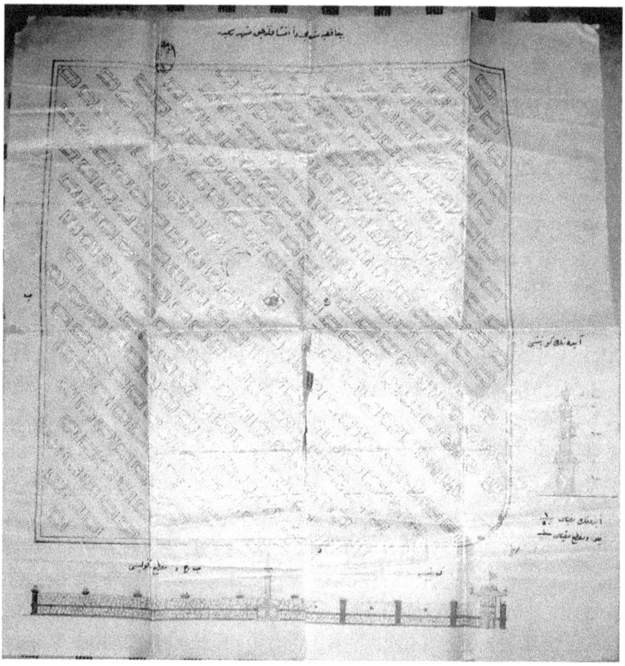

Figure 5.9 Namazgah military cemetery in Biga, on the Asian side of the Dardanelles.
Credit: General Directorate of State Archives of the Prime Ministry of the Republic of Turkey (BOA)

Figures 5.10(a) and (b) Detail of the commemorative monument in the Namazgah military cemetery in the Biga Municipality Cemetery (a), and its restored state today (b).

Credit: General Directorate of State Archives of the Prime Ministry of the Republic of Turkey (BOA)

soldiers—their efforts to establish and support the Ottoman state."[44] The actual plan of the cemetery, the paper of which is now tattered and torn along the folds of the document, shows well-organized parallel lines of grave-markers, an enclosing fence, and a sketch and measurements for the monument that would be included in the cemetery. This cemetery and its accompanying monument were built during the years of the Gallipoli battles, and still exist today in the Biga Municipality Cemetery, albeit in a much-restored state and without the adjacent hospital (Figures 5.10(a) and (b)).

In addition to planning for cemeteries and monuments to commemorate the Gallipoli dead during the war, commemorative days were also designated during the years of the war to honor the Gallipoli dead, and the dead of their wartime allies, the Germans.[45] Early in March 1916, the Ottoman government designated that 18 March, the date of the important Ottoman naval victory in the Dardanelles, would be recognized as the commemorative day for the Ottoman soldiers who died on the Gallipoli peninsula, and for Ottoman war veterans in general.[46] Initially Ottoman officials suggested that five days be established in the calendar year to commemorate those who had been killed during the battles on the peninsula. In addition to 18 March, these were the beginning of Ramazan, the birthdate and accession date of the reigning Ottoman sultan, and the birthday of the Prophet Mohammed.[47] Detailed instructions were issued and sent to the various military barracks and hospitals along the Dardanelles specifying the time, the place, the dress code, and the prayers that would be called for those who had died in battle. Commemorative activities were to begin on 18 March at 9:30 a.m. and be followed by ceremonies for the Ottoman soldiers who had died in battles on the peninsula or in hospital; an hour following that event, a ceremony would be held for the German dead.[48]

With the end of the War of Independence, various civic and government groups began campaigns to raise funds for veterans of the Gallipoli campaign, and for the construction of commemorative monuments on the peninsula.[49] As early as 30 October 1924, the Commission for the Construction of the Cemeteries of the Martyrs (*Şehitlik İmar Komisyonu*) was founded by the military to support the commemoration of veterans; following this, several veterans' societies were founded which included both government and civic groups and were tasked specifically for the purpose of raising funds and supervising the construction and upkeep of the cemeteries at Gallipoli.[50]

Figure 5.11 Atatürk on a visit in 1934 to the battlefields of the Çanakkale Wars with the Shah of Iran.

Credit: Suna and İnan Kıraç Foundation Photography Collection, IAE 0026278

The construction of a major commemorative monument on the Gallipoli peninsula was a central topic of discussion for the new Republican government of Turkey on 21 May 1926, and plans for one were proposed, and approved, the following year on 4 May 1927 in an order signed by Atatürk, the new president of the Republic.[51] In that document, the procedure for procuring supplies to build a monument dedicated to the martyrs of the Çanakkale Wars was specified. The monument would not be for any particular individual or specific group of soldiers, but would be dedicated to all who fought for and died on the Ottoman side, the identified and unknown martyrs of the war.

Atatürk appears to have been a frequent visitor to the peninsula during the post-war era and had visited Çanakkale five times between 1920 and 1935 (Figure 5.11). During his first trip to the battlefields since World War One on 1–2 Sept 1928, after becoming the president of Turkey, he walked the Anafartalar battlefields for four hours. This September visit was also noted by the British; in Lord Stopford's report about the famed 1928 St. Barnabas pilgrimage to Salonika and Gallipoli he mentions: "I was told that the Ghazi (President-Mustapha Kemal) was on the Peninsula and had been up to Lone Pine a couple of days previous; but we did not however see anything of him."[52]

Monuments to the "unknown soldier"

The first memorials to the "unknown soldier" appeared at different times in Europe after World War One.[53] In Britain and London, ceremonies for the unknown soldier took place two years after the end of the war, in November 1920. In Turkey, efforts to commemorate the dead and missing collectively with a tomb to an unknown soldier (*meçhul şehidi*) began as early as August 1925, not for the Gallipoli dead, but for soldiers who had died in the War of Independence in the central Anatolian town of Dumlupınar. The journalist Zekeriya Sertel made the suggestion in the periodical *Resimli Ay* that the new Turkish government should undertake the construction of a tomb for an unknown soldier.[54] It would commemorate the *Şehîd Sancakdâr*, or the unknown flag/standard bearer. The tomb of Dumlupınar, Sertel claims, would attract "national pilgrims" or "*milli hacılar.*"[55] According to Sertel, the dress of these pilgrims should not be the traditional white robe worn to Mecca, but the same red color as the blood of the martyrs who fought and died in the war. He envisioned that pilgrims would circumambulate the tomb of the unknown soldier as if it were the Kaaba.[56] The site of the unknown soldier would then become "a peak of victory which owes its existence to the accumulation of the bones of the martyrs . . . The hearts of all Turks are every day turned towards this direction of prayer (*kıble*) . . ."[57] The Dumlupınar memorial was completed by 1927.[58]

The establishment of a second memorial dedicated to the unknown soldier of the Balkan Wars was also discussed in 1927. It was intended to commemorate both the Ottoman soldiers and the civilians (around 10,000) who had been held as prisoners and perished in the Bulgarian concentration camps established during the Balkan Wars in the Sarayiçi district of Edirne. Rıfat Osman, an eyewitness to the tragedy that happened there, described the conditions of the camp in his memoirs and insisted that some kind of memorial should be erected for the many who died in these terrible conditions so as not to forget their sacrifices.[59] The proposed Sarayiçi Balkan Wars Memorial, also known as the Memorial for the Unknown Soldier (*Meçhul Asker Anıtı*), bears an inscription with the date of 26 March 1913 but was actually constructed on the shores of the Tunca River in Edirne in 1939, several decades after the Balkan Wars.[60]

Other commemorative monuments

By 1931 the British and French memorials and cemeteries on the Gallipoli peninsula had been completed and there was a significant increase in visitors and pilgrims to those graves, particularly from England and Australia. By this time, the comparatively few commemorative projects undertaken by the Turks on the peninsula were an increasing source of concern for several Turks. The need for a larger commemorative presence on the peninsula was voiced by some in the 1930s, such as Hüseyin Atsız, who was vehemently against the presence of foreign and Christian monuments on the peninsula and urged the youth of Turkey to take a stand against them: "Enemy monuments in your homeland . . . Turkish youth, how are you enduring this? . . . why do you not destroy these infidel monuments?"[61] For Atsız, like Sertel, the Gallipoli peninsula and other areas where Turks had fought for their homelands should be considered as the Meccas of new Turkey, and it was a duty of all Turks to make pilgrimages there as a test of their manhood.[62] A member of parliament from Edirne pointed out in 1931 that there was still little on the Gallipoli peninsula that could serve as a tangible reminder of the sacrifices made by the Ottoman soldiers who had lost their lives fighting there. This, he remarked, contrasted poorly with the British and French commemorative memorials in their newly completed and pristine condition.[63]

The comparatively fewer examples of commemorative architecture on the peninsula created to honor the Gallipoli dead on the Ottoman side, along with the basic styles and rudimentary materials used in these structures, do need to be assessed along with the state of the architectural profession in Turkey at the time, and the economy in these immediate post-war years. Increasingly in the architectural profession there was a desire to jettison the neo-Ottoman revivalist style which had been popular in the final decade of the empire. [64] While Neo-Ottoman architects had initially favored bringing together elements considered of Ottoman origin with the technology and design principles of European civilization, by 1931 the architects of the newly established Turkish Republic were turning to the unadorned stripped-down aesthetics of German and Central European modernism.[65] The result was the New Architecture—the "*Yeni Mimari*"—and it represented a break, at least in its rhetoric, with the Ottoman past.[66]

One of the earliest designs for a more substantial memorial on the Gallipoli peninsula reflecting this new passion for the modern was proposed by Şahin Naci, the drawing of which was published in a 1933 issue of *Yedigün*, a widely circulated Turkish newspaper.[67] Şahin Naci's "March 5th" monument was intended to commemorate an Ottoman victory during the Gallipoli campaign at Anafartalar, and was a good example of the new aesthetics of the Yeni Mimar movement.

While it was never constructed, from the sketch of the monument published in *Yedigün* it appears to have been inspired almost entirely by the modernist trends prevalent in Europe at the time; there was little in its design that could be identified as uniquely Ottoman or Turkish. The structure was to be a central column of 30 meters which supported an illuminated crystal dome. This was to be flanked by two supportive piers, each with a large urn where flames could continually burn. Panels of the monument's façade were to be inscribed with the names of the different regiments who had fought and died in the Anafartalar battles (Figures 5.12(a) and (b)).

Contrary to the practice of commemoration that had emerged with World War One monuments in Europe, or memorials such as the Menin Gate in Ypres and the CWGC memorials in Gallipoli, the individual names of the dead or missing soldiers were not to appear on the Turkish monument. Equality in death by the end of World War One had become the standard commemorative practice in Europe and the United States. As Inglis and others have pointed out, "in most countries the names of the listed on the memorials are usually given in alphabetical order, not in order of rank."[68] This practice was not, however, followed in the new Turkish Republic and hierarchy continued to shape the identities of the Gallipoli dead as it had shaped the soldiers' lives before they passed away. There were other factors influencing the decision not to list the names of the individual dead on the memorials, among these the fact that listing last names alphabetically would have posed problems in Turkey prior to 1934, the year that the Turkish government passed a Surname Law requiring that all citizens of Turkey adopt a

Figures 5.12(a) and (b) Şahin Naci's "March 5th" monument.

Credit: *Yedigün*, 1933

family name, and not use Ottoman titles such as Pasha, or an extension of their father's name, or a nickname.[69]

It is not clear why Şahin Naci Bey's March 5th monument was never constructed. In 1934, the year following the publication of his drawing, the Turkish Ministry of Public Works passed legislation which stated that it was of great importance to develop an architectural style that was more Turkish. Prominent young Turkish architects working at the time, such as Sedad Hakkı Eldem, began promoting an architectural aesthetic which brought together elements of "native" and vernacular Turkish architecture e.g. Turkish house, and the bare aesthetics of European modernist buildings. Eldem, his students, and many of the architects of the time were teaching, designing, and disseminating ideas about architecture by publishing their drawings in journals such as *Arkitekt*, and in other forums dedicated to promoting "Turkishness" and modernity in the New Republic. As Bozdoğan and Akcan note, "these years marked the beginning of an ideologically charged binary way of thinking in terms of contrasts between the old and the new, the traditional and the contemporary (*asri*), the reactionary and the progressive, which permeated the entire architectural discourse of the early republic."[70]

Labor shortages and indecision over how to integrate the modernist styles of Europe were not the only challenges impacting the progress of commemorative architectural projects on the Gallipoli peninsula. The state's financial difficulties during the post-WWI economic depression also contributed to a delay in the design and construction of memorial projects there. Even though an association called the *Çanakkale Şehitleri Abidelerine Yardım Derneği* (Association to Support the Memorials of the Çanakkale Martyrs) was founded in 1938 to raise funds to build a monument on the peninsula, sufficient resources could not be amassed and no substantive steps were taken to initiate a project for another decade.

The death of Atatürk on 10 November 1938 also shifted whatever resources were available for national commemorative projects away from Gallipoli to the capital of Ankara and the creation of the *Anıtkabir*, the commemorative structure built for the nation's founder. Between 1938 and1953, the construction of this massive commemorative monument in Ankara captured the interest and energy of many Turkish and foreign architects working in the country at the time. Within a month of the death of Atatürk, a committee had been formed to decide on the location of the monument. As Atatürk had been a formidable figure in the Çanakkale Wars and was also the hero of the Turkish War of Independence, any monument to commemorate him was also intended to be more than a memorial for an individual hero; rather, it would symbolize collectively all who had made sacrifices for the entire nation. As Wilson has noted, "The memories of both the man Mustafa Kemal Atatürk and the memories of the nation Turkey are tied together in the funerary architecture for Atatürk, as politicized by the successive governments and citizens of Turkey since Atatürk's death in 1938."[71]

By 1 March 1941 an international competition for the design of the commemorative complex in Ankara had been announced. Initially, and somewhat surprisingly, only foreign architects were going to be allowed to submit designs for the monument. It was assumed by the selection committee, comprising several

Figure 5.13 The winning design of the Ankara Anıtkabir by Onat and Arda.
Credit: *Arkitekt*, 1943

Turkish members of parliament and ministers as well as foreign architects who had been commissioned to design many of the government buildings for the nation's new capital in Ankara, that non-Turkish architects could best realize the kind of monumentality for an edifice that would properly remember the nation's founder. Protests from the Turkish Chamber of Architects succeeded in opening this competition and other state commissions to Turkish architects; among the 49 entries for Atatürk's mausoleum, half were submitted by Turkish architects.[72]

The prize-winning drawings of Emin Onat and Orhan Arda from the architectural competition for the Anıtkabir were published in 1943 in *Arkitekt*, the leading Turkish architectural journal at the time (Figure 5.13).[73] The wide range of designs submitted by both foreign and Turkish architects for the competition were also displayed to the public during the week of 24–31 March 1942 at the Ankara Exhibition Building, so they would have been easily available and served as a source of inspiration for the architects who would participate shortly thereafter in the competition for a memorial on the Gallipoli peninsula. Indeed, some of the architects who participated in the Ankara Anıtkabir competition went on to submit entries for the Çanakkale Martyrs Memorial competition, among them Sedad Hakkı Eldem and his team, who received an honorable mention.[74] But only after the government completed the process of erecting the Anıtkabir in Ankara could it turn its attentions and finances to a second national and major commemorative project on the Gallipoli peninsula.

A new memorial for Gallipoli: the Çanakkale Martyrs Memorial

Only Turks were allowed to join the 1944 *Çanakkale Zaferi ve Meçhul Asker Anıtı Yarışması*, the Competition for the Monument to the Victory of Çanakkale and the Unknown Soldier, the first major project to commemorate the collective Turkish dead and missing on the Gallipoli peninsula. Thirty-seven entries were submitted, of which one had to be eliminated due to a technical problem with its submission documents. The remaining 36 entries were reviewed by the jury, which met from 15 to 19 March 1944 to determine the winning project. The collective decisions of the jury and a selection of the drawings of the winning project were published in *Arkitekt* that same year, along with some drawings from the projects which received second, third, and honorable mentions.[75]

From the comments made by the jury about several of the submissions for the Çanakkale memorial it appears there was little tolerance for monuments with exterior decoration, and not much sympathy for designs that evoked any kind of Ottoman or pre-Ottoman Selçuk architecture; these were now seen as representing a failed empire and outdated styles. Lessons had been learned from the winning design of the Ankara Anıtkabir and some of the submissions seem to have taken their cues from Emin Onat and Orhan Arda's First Prize plans for the Ankara mausoleum. [76]

Sedad Hakkı Eldem's entries for both the Ankara and Gallipoli competitions exemplify the stylistic shift that was occurring in the 1930–40s among Turkish architects. While his submitted design for the Anıtkabir was inspired by a *kümbet*, or Selçuk tomb tower, and other examples of traditional architectural forms in Anatolia, his entry for the Çanakkale memorial was a massive and unadorned baldachin structure with a central oculus supported by four large arches, and had few prototypes in Anatolia (Figures 5.14(a) and (b)).[77]

The jury for the Çanakkale Memorial and the competitors in this second major commemorative project clearly favored the international modernist style of Europe rather than a synthesis of this style with the vernacular or historical architecture of Turkey. What won in Çanakkale was an immense square baldachin structure which exhibits far less of the tensions seen in the projects submitted for the Anıtkabir, as it is securely rooted in the international modernist style of the 1940s (Figure 5.15). It was also the least expensive option, coming in at half the price of all the other entries.[78] The jury praised İsmail Utkular, Doğan Erginbaş, and Feridun Kip's winning proposal for its simplicity, grandeur, and the fact that it was easy to construct and cost-effective.[79] Reflecting the austere economics of the time, several of the submissions that lost the competition were eliminated for being too decorative and/or just too costly to build. Project #37, for example, lost the competition and was criticized by the jury for being the most expensive and the most decorated of all the entries submitted.[80]

The monument in Çanakkale would be made of concrete and covered with granite. Initially it was to be located at Alçıtepe, one of the higher points in the interior of the peninsula's southern section, but both the lack of space there and the unfavorable results of the initial geodesy research forced the committee to look at the current location of the monument, referred to today as the *Abide*, on the promontory just to the north of Cape Helles, at Hisarlık, the site of ancient Elaious. Neither the architects who had won the competition, nor the planning board in charge of its execution were concerned that the great commemorative

Figures 5.14(a) and (b) Sedad Hakkı Eldem's Selçuk tomb design for the Anıtkabir competition (a) differed significantly from the more classical entry for the Çanakkale competition (b).

Credit: *Arkitekt*, 1943 and 1944.

monument, with its foundations burrowing 7.5 meters into the earth and 625 square meters of grounds spreading out around the structure, was located on the important Antique site which had been excavated by the French two decades earlier.[44] No new archaeological survey was proposed or conducted of Elaious before the digging for the massive foundations of the great baldachin memorial structure began.[81]

Concerns that the jury had expressed about the costs of the monument proposed for the Abide proved to be valid as difficulties were encountered early on in its financing and the project was put on hold for eight years. Renewed interest in the Abide project emerged in 1952, as the Anıtkabir's construction in Ankara was

Figure 15.5 The winning design in the 1944 competition for the Çanakkale memorial.
Credit: *Arkitekt*, 1944

coming to an end. Financing Atatürk's memorial in the capital had been a challenge and a drain on the nation's finances and cost-cutting revisions had been made several times during the construction process. The Ankara memorial was, at last, completed and ready for its opening on the 15th anniversary of the death of Atatürk, 10 November 1953.[82] With the Anıtkabir completed, attention could at last turn to the Gallipoli peninsula—but this project proved to be even more problematic and difficult to finance than the Anıtkabir in Ankara.

Efforts to gather funds from the public for the Gallipoli monument were made by civic groups such as Society for the Assistance for the Construction of the Çanakkale Martyrs Memorial (*Şehitlikleri İmar Cemiyeti Çanakkale Şehitler Abidesi İnşaatına Yardım Komitesi*), which had been established in 1953.[83] In March 1954, the construction tender was won by an engineer, Sırrı Onural, and timber supplier Hayrettin Aksoy for 1,648,234 Turkish lira. Bureaucratic formalities were resolved; celebrations were held between 17–19 April 1954 to mark the first work on the foundations of the monument. The construction was slated to take 500 days.

Troubles began soon after the celebrations ended. The technical committee in charge of inspecting the monument's initial concrete foundation discovered that the contractors had made these improperly; at least 40 tons of poor grade iron was deemed unusable and other construction materials which had been purchased on the black market were also found to be of inferior quality. Legal action was taken against the contractors; in the interim all building activity was halted. It would not resume until 1956 when a new tender was announced in the amount of 1,708,918 Turkish lira. This time a team of engineers—Ertuğrul Bana, Osman Gencal, and Muammer Yeşildağ—promised to complete the monument by July 1958, which at that point looked possible; by 1957 the structure was wrapped in scaffolding, and had risen to 12 of its anticipated 42 meters.

However, once again progress towards completing the memorial was set back by financial mismanagement. Due to the impact on the budget of the initial corruption and the rising cost of building materials, funds ran out again and construction ceased entirely on 15 January 1958. The future of the monument at this point looked quite bleak. To determine the fate of the Abide, yet another committee was appointed.[84] This one opted go to the public and announce the dire situation to some of the more popular newspapers in the country, such as *Milliyet*.[85] With the government unable to supply any more funds, the committee hoped to solicit from the public the remaining funds needed to complete the project, an estimated 900,000 lira.

The results were astoundingly successful, and the example of civic participation to support a commemorative project was a landmark event in Turkey. Turks from across the social spectrum contributed personal funds. Donations included "free shifts offered by construction workers, blankets with monument motives knitted by women in Siirt, charity concerts given by a Jewish singer, Dario Moreno, and a Monument Cup organized among football clubs."[86] One member of parliament allocated his entire annual salary to the project; artists gave concerts to raise money, Turkish school groups, artists' societies, women's groups, and many others contributed, and at last a total of 1,686,251 lira was collected. The Istanbul municipality gave an additional 100,000 lira and the National Assembly contributed 600,000 lira. In total, 4,153,761 lira was raised.[87] With this new infusion of cash the basic shell of the monument was completed, but several parts of the initial design for the project—including the harbor, the infrastructure for illuminating the monument at night, a museum of wartime memorabilia, and a new entrance road—had to be abandoned, prompting the criticism by Onur who noted that the monument was still severely lacking its finishing touches.[88] At last, on 21 August 1960, the Çanakkale Martyrs Memorial was officially opened. Work on the exterior continued, however, and it was only at the end of July 1999 that eight relief carvings which had been originally planned for the façades of the monument's four supporting columns were finished. With the monument reaching a height of just over 40 meters, it became the most visible landmark in the region, especially today as it is illuminated with red floodlights at night.[89] It was declared a cultural landmark by the Turkish Ministry of Culture and Tourism in 1980 (Figure 5.16).

The commonly stated explanation that the Turks did not care for their dead and missing on the Gallipoli peninsula until 1944, when the first plans for the Çanakkale Martyrs Memorial began to take shape, is therefore not accurate. Of the commemorative monuments constructed by the Ottomans during and shortly after the war on the peninsula, some were destroyed by Australians and French during the years prior to the Treaty of Lausanne when the ownership status of the Gallipoli peninsula was uncertain. Efforts made throughout the 1930s to establish a greater

Figure 5.16 The Çanakkale Martyrs Memorial today.
Credit: Archive of L. Thys-Şenocak

commemorative presence on the peninsula met with limited success due to financial problems facing the new Republic, and the planning process for the Abide in Gallipoli, in terms of both its design and contruction, were impacted by the great commemorative project of the Anıtkabir of Ankara, which consumed the resources of the government from 1942 to 1953.

The Anıtkabir in Ankara is a funerary monument to an individual; it also serves to commemorate the birth of the Turkish nation and those who, like Atatürk, had fought or sacrificed their lives defending that nation in the battles on the Gallipoli peninsula, and in other places and other wars such as the War of Independence. In this sense, the need to complete an additional commemorative monument on the Gallipoli peninsula became less urgent than the creation of a memorial to commemorate the death of the founder of the nation in the new capital of that nation. Understanding the chronologies of these two major commemorative projects in Ankara and the Gallipoli peninsula is essential, as it is clear that when resources permitted, and attentions and funding shifted away from the Ankara project, the planning and construction of a major commemorative monument to the Ottoman martyrs of the Gallipoli battles did proceed, and was enthusiastically supported by the public who partially financed it.[90] With the completion of the Çanakkale Martyrs Memorial, Turkey clearly proclaimed its ownership of the Gallipoli peninsula through the production of this mammoth edifice. While it was not the first commemorative structure to have been built on the peninsula by the Turks, it was by far the most impressive memorial that had been erected there up to that point. It also heralded a profusion of building projects that would be undertaken by the Turkish government, for both commemorative and touristic purposes.

Notes

1 Cumhuriyet, 30 Ağustos, 1933; Mithat Atabay et al., "Remembering Gallipoli from a Turkish Perspective," in *Anzac Battlefield*, 238.
2 Ibid.
3 Ibid., 236.
4 Ibid., 238.
5 For a discussion of this see Z. Türkmen, A. Çalışkan and F. İlhan, *Çanakkale Savaşı Anıtları ve Şehitlikleri (The Monuments and Martyrs' Graves of the Çanakkale War)* (Ankara: Genelkurmay Basımevi, 2007).
6 Alan Moorehead, *Gallipoli 1915* (New York: Ballantine Books, 1956) 314.
7 J. Laffin, "Gallipoli the Largest British War Memorial in the World", *Royal British Legion Journal*, March, 1974; C.E. Bean had noted that during the cease-fire on 24 May 1915 the Turks ". . . sometimes collected the bodies in the nearest washaway that scored the steep gully sides, and shoveled the banks of it over them." See his *First World Official War Histories*, 1st edition, 1948, p. 58.
8 CWGC 9/928/G/V.
9 Pemberton, *To-Day*, 20.
10 Ibid., 38.
11 C.E. Bean, *First World War Official Histories*, 1st edition, 1948, Chapter 4, 48–9; Pemberton, 37–8.

12 R. Bademli, K. Sarı et al., *Gelibolu Yarımadası Tarihi Milli Parkı Uzun Devreli Gelişme Planı Çalışmaları (The Working Report of the National Peace Park: Long-Term Development Plan for the Gallipoli Peninsula (UDGP)* (Ankara: Middle East Technical University, 2002), 84.

13 Sibel Baykut, "The Re-contextualization of the Battle of Gallipoli through Commemorations," *Intercultural Understanding*, 6 (2016), 9.

14 Pemberton, 38.

15 Afet İnan, *Atatürk Hakkında Hatıralar ve Belgeler* (İstanbul: Türkiye İş Bankası Kültür Yayınları, 1984), 217.

16 Jay Winter, *Sites of Memory, Sites of Mourning*, 80, 94.

17 Edhem Eldem, "State, Nation and Death," *Death in Istanbul: Death and its Rituals in Ottoman-Islamic Culture* (İstanbul: Ottoman Bank Archive and Research Centre, 2005), 270.

18 Ibid.

19 The four-columned monument in Haifa which in 1902 commemorated Sultan Abdülhamid II's construction of the railway from Damascus to Mecca, the Hamidiye Hijaz line, is an example of the former; several of the commemorative monuments in Edirne which commemorate the dead from the Balkan Wars are examples of early obelisk-type memorials.

20 Klaus Kreiser, "Public Monuments in Turkey and Egypt 1840–1916," *Muqarnas* XIV (1997), 103.

21 Alan Borg, *War Memorials from Antiquity to the Present* (London: Leo Cooper, 1991), 70.

22 Eldem, 270.

23 Metin Sözen, Türklerde Anıt, *Mimarlık*, 7 (1973), 10.

24 Kreiser, 113.

25 Ibid., 103–4.

26 Until 1935 all pilots in Turkey would pay their respects to these fallen aviators by observing a one-hour suspension of flights throughout the country each 27 January. After 1935 this date was changed to 19 May, and then eliminated in 2002 when all veterans were recognized on the same date: 18 March. See www.tayyareci.com/hvsehitleri/hvsegunu.asp.

27 Eldem, 276.

28 The Monument to the Gendarmerie *(Jandarma Abidesi)*, now in Karaağaç, Bulgaria, is an example of this early collective memorial which was constructed by the state for the Ottoman dead from the First Balkan War and was built with a "European style" (*Avrupa tarzında*). See Oral Onur and Latif Boğman, *Edirne Şehit Anıtları* (Türk Şehitlikleri İmar Vakfı Yayınları no. 2, 2012), 21. For example, the "*Babadan Oğula Anıtı"* (Father to Son Memorial) built during the Balkan Wars in the village of Arnavut, Edirne gives the name of the colonel but 15 other soldiers and officers of different ranks are interred here as well. For information on the Balkan Wars see Syed Tanvır Wasti, "The 1912–13 Balkan Wars and the Siege of Edirne," *Middle Eastern Studies* 40, no. 4 (2004).

29 Onur and Boğman, *Edirne Şehit Anıtları*, 28–9; for the *Söğütlük* memorial see 63–9.

30 Ibid., 63.

31 Ibid. 63–70.

32 For a study of these monuments and the proposal to reconstruct several see Mithat Atabay, *Gelibolu Yarımadası'nda Yazılan Tarih Yıkılan Anıtlar* (İstanbul: Paradigma Akademi, 2017).

33 Atabay et al., "Remembering Gallipoli from a Turkish Perspective," *Anzac Battlefield*, 233–8.

34 Ibid.

35 C.E. Bean, *First World War Official Histories*, 48–9.

36 C.E. Bean in *First World War Official Histories*, notes the presence of these graves, the monument on north beach, Çataldere and Lone Pine cemeteries, 49, 60–6. Atabay et al., *Anzac Battlefield*, 237–8. For the Anafartalar Victory Monument see Mithat Atabay, *Gelibolu Yarımadası'nda yazılan Tarih Yıkılan Anıtlar*, 73–9.

37 A proposal for the reconstruction of this monument has been proposed by M. Atabay but the current restoration plan for Seddülbahir fortress does not include the reconstruction of this monument.

38 Ahenk Yılmaz, "Architectural Memorialization of War: Ars Memoriae and the Landscape of the Gallipoli Battles", Ph.D. diss., (Izmir Technical University, 2014), 45.

39 See G7432.G1 S65VIII.27/34 in the Australian War Memorial for the Şevki Pasha maps.

40 Mehmet Şevki Paşa, *Çanakkale Tahkimat Haritası/Defense Atlas of Çanakkale* (Ankara: Genelkurmay Askeri Tarih ve Stratejik Etüt Başkanlığı Yayınları, Genelkurmay Basımevi, 2009); Murat Karataş, *Haritalarla Çanakkale Savaşları: Gelibolu Yarımadası Kuzey Bölgesi Kara Muharebeleri/Maps of the Çanakkale Wars: Battles in the North Region of the Gallipoli Peninsula* (Ankara: Nobel, 2007). See also www.awm.gov.au/exhibitions/gmaps/turkish/ANZAC.asp.

41 Several of the locations on Şevki Pasha's map were rediscovered during the archaeological survey and documentation project of the ANZAC area discussed in Chapter 3 that was initiated in 2005 by the geophysical engineering department of 18 March University in Çanakkale, the Directorate of the National Park, and the University of Melbourne.

42 The project to identify the names and places of origins of the Ottoman martyrs and compile these into a database was called the TMP Türk Şehitlık ve Anıtlar Envanteri. Information about the different types of symbolic and real cemeteries can be found at: https://ipfs.io/ipfs/QmT5NvUtoM5nWFfrQdVrFtvGfKFmG7AHE8P34isapyhCxX/wiki/Gelibolu_Yar%C4%B1madas%C4%B1_Tarihi_Milli_Park%C4%B1_%C5%9Fehitlikleri_ve_eserleri.html. ÇATAB currently provides on its website the Ministry of Defense's list of all known Çanakkale martyrs. See http://catab.kulturturizm.gov.tr/Eklenti/48077,msb-11042016pdf.pdf?0&_tag1=60358D047E9F11FAB4A78E1EE1E5797F9C79BE84. The names of Çanakkale martyrs are also listed under the burial site. http://catab.kulturturizm.gov.tr/TR,163324/canakkale-sehitleri.html.

43 Mithat Atabay, M. Erat and H. Çobanoğlu, *Çanakkale Şehitlikleri* (İstanbul: İş Bankası Kültür Yayınları, 2009), 66–82, 102–27. The İntepe Martyrs memorial would be constructed off the peninsula on the Asian side of the Dardanelles in 1934.

44 BOA (Başbakanlık Osmanlı Arşivi, DH.UMVM (Dahiliye Nezareti Umur-u Mahalliye ve Vilayat Müdüriyeti Evrak Vi137/24), (19 March 1916).

45 There are several recent studies of the chronology of the establishment of commemorative days and events associated with the Çanakkale War, particularly 18 March. For detailed information on these events and the significance of 18 March in the past century see J. Macleod, *Gallipoli*, 154–87; Kant, "Çanakkale's Children," 145–64; M. Atabay et al., *Anzac Battlefield*, 222–43; K. Şakul, "Contemporary Turkish Perceptions of the Gallipoli Campaign," *The Gallipoli Campaign: The Turkish Perspective*, ed. Metin Gürcan and Robert Johnson (London and New York: Routledge,

2016), 181–204; Sibel Baykut, "The Re-contextualization of the Battle of Gallipoli through Commemorations," *Intercultural Understanding*, vol. 6 (2016), 7–15.

46 This date was decided by the Turkish Parliament to be National Martyrs Day in June 2002. Kant, 153.

47 Genelkurmay Başkanlığı Askeri Tarih ve Stratejik Etüt Başkanlığı Arşivi, *Birinci Dünya Savaşı Koleksiyonu*, Klasör: 4685, Dosya: 340, Fihrist: 45. 5 Mart 1331/ (18 Mart 1915).

48 Ibid.

49 Some "monuments" dedicated to the support of the Gallipoli front had emerged in Istanbul during the war, such as *Hatıra-i Celadet Topu*, a monument constructed in the tradition of the "nail monuments" of Germanic origin. This was a cannon fabricated of wood and located in Bayezid Square, Istanbul, to which those who wanted to support the war efforts in Çanakkale could do so by adding nails of different metallic values to the cannon exterior. See Yiğit Akın, "Savaş, Milliyetçilik, Sivil Toplum: Çivi Anıtların İstanbul'daki Örneği Hatıra-i Celadet Topu," *Toplumsal Tarih* (March 2014), 62–7; Winter, *Sites of Memory*, 82–5 also discusses the nail memorials as a German and Austrian practice and not something found in France or Britain.

50 Onur and Boğman, 4.

51 Turkish Republican Archives Ankara (BCA) "Çanakkale Şehitlik Abidesi'nin emaneten yaptırılması," 4 May 1927.

52 "St. Barnabas Pilgrimage to Salonika and Gallipoli 1928," CWGC 1/1/16/25 and CWGC 1/1/16/26.

53 K.S. Inglis, "War Memorials: Ten Questions for Historians," *Guerres mondiales et conflits contemporains, No. 167, Les Monuments Aux Morts de la Premiere Guerre Mondiale* (Paris: Presses Universitaires de France, July 1992), 5–21.

54 Klaus Kreiser, "Public Monuments in Kemalist and Post-Kemalist Turkey," *Türklük Bilgisi Araştırmaları/Journal of Turkish Studies* 26/11 (2002), 43–60, 47; See also Zekeriya Sertel, *Resimli Ay*, (Ağustos/August 1925), 25.

55 "*hacı"* is the term used by Turks to describe those who make the pilgrimage to the Kaaba in Mecca as part of the duties of Islam.

56 Kreiser, "Public Monuments," 47.

57 *Sertel,* 25.

58 Bedrettin Tuncel, *Atatürk ve 30 Ağustos Zaferi'nin İlk Kutlanışı* (Ankara: TTK, 1972).

59 For the observations of Rıfat Osman (Tosyavizade) on Edirne see *Edirne Abideleri Yazıları.* ed R. Duymaz and Y. Topaloğlu, (Edirne: Trakya University, 2015).

60 Onur and Boğman, 72–6. The project was designed by the architect Kemal Altan who was the architect for the Undersecretariat of the Thracian Inspectors (Trakya Müfettişliği Bayındırlık Müşaviri).

61 Hüseyin Nihal Atsız, *Çanakkale'ye Yürüyüş* (İstanbul, 1933) republished (İstanbul: İrfan Publishing, 2015), 5–7, 13.

62 Ibid., 3.

63 Turkish Republican Archives Ankara (BCA). 030.010.00.00.3.18.10.3–4 (Anafartalar' daki Mehmet Çavuş Şehitliği'nin imarı hakkında Edirne Mebusu Şeref Bey'in kanun teklifi. Mebusu Şeref Bey, 14 November 1931).

64 Sibel Bozdoğan, *Modernism and Nation Building: Turkish Architectural Culture in the Early Republic* (Seattle: University of Washington Press, 2001), 33.

65 Ibid., 47.

66 Ibid.

67 *Yedigün*, (November 1933 no. 34), 14.

68 Inglis, 6–7.
69 For an overview of the Surname Law in Turkey see: Meltem Türköz, "Surname Narratives and the State-Society Boundary: Memories of Turkey's Family Name Law of 1934," *Middle Eastern Studies*, vol. 43, no. 6 (2007), 893–908.
70 Sibel Bozdoğan and Esra Akcan, *Turkey: Modern Architectures in History* (London: Reaktion Books, 2012), 24.
71 Christopher Wilson, *Beyond Anıtkabir: The Funerary Architecture of Atatürk* (Farnham, Surrey: Ashgate, 2013), 130.
72 The exact number of entries is unclear. Wilson states that there were 50 entries for the competition (71); Bozdoğan Modernism and Nation Building, (286) claims there were 49 entries.
73 Z. Sayar, "Anıt-Kabir Müsabakası münasebetiyle,"*Arkitekt*, no. 3–4 (1943), 5.
74 "Çanakkale Zaferi ve Meçhul Asker Abidesi Müsabakası (I. Ödül)," *Arkitekt*, no. 3–4 (1944), 59–60.
75 "Çanakkale Zaferi ve Meçhul Asker Abidesi Müsabakası (I. Ödül)," Arkitekt no. 3–4 (1944), 52–74.
76 Emin Onat and Orhan Arda, "Anıt-Kabir" *Arkitekt* Issue 280, no. 2 (1955), 55–9; Bozdoğan, 289.
77 *Arkitekt*, İstanbul 1944, no. 3–4, 57–8.
78 *Arkitekt*, İstanbul 1944, no. 3–4, 65. "İnşası kolay ve mümkündür. Diğer projelere nisbetle masrafı %50' dir."
79 Ibid.
80 *Arkitekt* İstanbul 1944, Issue 3–4, 74. ("En büyük, en masraflı bir projedir. Fazla dekoratiftir.")
81 Interview with Ertuğrul Ceylan, 27 November 2015. Koç University Oral History and Memory Archive, Thys-Şenocak collection, Koç University Suna Kıraç Library (http://digitalcollections.library.ku.edu.tr/cdm/).
82 Wilson, 118. Wilson has estimated that at this time 10 million Turkish lira would be the equivalent in 2013 of 90 million US dollars.
83 Necmi Onur, *Çanakkale Savaşları ve Şehitler Abidesi* (İstanbul: Gün Matbaası, 1960).
84 The committee members were Nihat Sözler, Hikmet Arda, Ihsan Üner, Cevdet Bereket, Coşkun Meltem, Mükerrem Taşçıoğlu, İffet Halim Oruz and Ertuğrul Ceylan.
85 Interview with Ertuğrul Ceylan on 27 November 2015. See the following issues of the newspaper *Milliyet* for information about the progress of the Abide project: 5 May 1952; 3 July 1952; 9 July 1952; 26 July 1952; 29 July 1952; 10 August 1952; 18 January 1958; 19 January 1958; 20 January 1958; 6 April 1958; 21 March 1963.
86 Sibel Baykut, "The Re-contextualization of the Battle of Gallipoli through Commemorations," *Intercultural Understanding*, vol. 6 (2016), 9.
87 "Abide için yardım 1.5 milyon aştı," *Milliyet*, 6 April 1958, 1.
88 Necmi Onur, "Şehitler abidesi alaka bekliyor," *Milliyet*, 21 March 1963, 3.
89 Baykut, 10. Baykut adds the popular belief that, "The height of the monument is of great significance as 40 meters symbolizes the fallen soldiers at Gallipoli and the remaining 1.70 meters symbolizes Mustafa Kemal's own height."
90 The technical committee for the Abide included one of the winners of the Ankara Anıtkabir project: Emin Onat.

6 The future of the Gallipoli peninsula

Towards 2023

According to the Çanakkale Wars and Gallipoli Historic Region Directorate (ÇATAB), the organization which has been tasked with the administration and long-term planning for the Gallipoli peninsula since 2014, its mission is to:

> "Develop as an open-air museum the Gelibolu Historical Area, where we paid the price during the Çanakkale Wars for our national unity and togetherness; to protect this historical area's emotional, historical, and cultural values, along with its nature, using a universal perspective, a sense of responsibility for the future generations, and cooperation with the veteran villages."[1] In its vision statement, ÇATAB promises that " 'The Spirit of Çanakkale' which is the symbol of our national unity and togetherness, will exist in the future as it has in the past."[2]

Both the mission and vision statements of ÇATAB, along with its 2016–2020 strategic planning document, provide important insights into how ÇATAB has conceptualized past efforts and commemorative projects undertaken on the Gallipoli peninsula by different governments, national and foreign—and more importantly, how new narratives will be shaped and projects implemented in the future to preserve and promote both the tangible and intangible heritage of Gallipoli.[3] As of July 2017, ÇATAB counts as tangible national heritage on the peninsula 50 Turkish martyrs' cemeteries, 29 Turkish memorials and epigraphic monuments, 34 cemeteries and memorials, 6 fortresses, 14 bastions, 32 archaeological sites, and 36 underwater wrecks. There are 8 villages in the area managed by ÇATAB and in 2015 the total population in these villages was 8,188 people.[4]

Competing narratives

Previous chapters in this book have pointed out that several competing heritage narratives have taken root on the Gallipoli peninsula since the end of World War One and most of these have emphasized different national and international political agendas. The story of the emergence of Australian and Turkish nationhood has forefronted the role that the battles of Gallipoli played—and continue to play—in the identity formation of these two countries. The position of Gallipoli in the history of the "coming of age" of Australia and New Zealand, and the nationalist pilgrimage route that has developed since the 1920s on the shores of this distant battlefield in the Aegean, are thoroughly studied subjects and do not need to be elaborated here.[5] Building on the individual nation-identity formation discourses of Turkey and Australia is the powerful story of camaraderie between these foes during and after the battle. This particular narrative has been inscribed onto the landscape of Gallipoli for more than seven decades and continues to flourish. The linked and common fate of unfortunate soldiers—the "Johnnies and Mehmets," young men who were sacrificed by the greater imperial powers of that age—is a useful cornerstone of diplomacy for all countries involved. As noted in Chapter One, the poetic dimension of that particular narrative, which presents the Gallipoli battles as a "gentleman's war," was included in the documents submitted by the Turkish government to UNESCO in 2014 for nomination to the tentative list of World Heritage sites.

Another powerful narrative inscribed onto the landscape of Gallipoli has been shaped by the post-WWI secularist citizens of Turkey, and has focused on the important role played in the battles by the nation's founder Mustafa Kemal Pasha: Atatürk. As Okur and others have noted, "The Dardanelles Campaign was among the national myths which, it was hoped, would function as the psychological building blocks of a 'new' Turkishness."[6] Proponents of this narrative associate the landscape of Gallipoli primarily with the heroism and tenacity demonstrated by the nation's founder, a key military figure in the Gallipoli battles. Other aspects of Atatürk's later, post-WWI achievements, including the victory of the Turkish War of Independence, are also a part of this narrative of struggle and success. The continuity between the Ottoman victory at Gallipoli and the successes of Atatürk in the Turkish War of Independence has been nurtured by the Turkish government when advantageous, as this telling of the past leaves out the final defeat of the Ottoman Empire in World War One, and the subsequent Allied occupation of both the Dardanelles and Constantinople from 1918 to 1923. Some scholars have noted that the foundation myths constructed about the birth of the Turkish Republic so closely link the battles fought on the Gallipoli peninsula with the victories of the Turkish War of Independence that many Turks today "often mistake it as a part of the War of Independence."[7]

With the victory in the 2002 elections in Turkey of the religiously conservative Justice and Development (*Adalet ve Kalkınma* or AK) party, the more secularist narrative about state formation that had been associated with the Gallipoli landscape in the preceding decades began to change. Veysel Eroğlu, the Minister of Forestry and Water Affairs, describing a visit by former Prime Minister (and now President) Erdoğan to the peninsula on 18 March 2003, reported that the latter gave orders to "save the martyrs' cemeteries in Gallipoli from the state of neglect."[8] This was among the catalysts for several projects to begin on the peninsula and spurred the creation of a new heritage narrative for this landscape. More buildings to accommodate tourists were constructed by the Turkish Ministry of Forestry and Water Affairs in 2005, such as the Kilye visitors' center at the entrance to the park. In addition, with the help of the Ottoman-era Şevki Pasha maps of the post-WWI landscape of Gallipoli, several mass graves belonging to Ottoman soldiers were identified via non-invasive archaeological methods such as geo-radar; new commemorative sites were created, and cemeteries such as the Ağadere Hospital and Memorial in Kilitbahir were prepared for visitors. Along with these developments came more buildings which could serve increasing numbers of tourists, larger parking lots, and wider roads for tourist buses to follow the various routes around the peninsula.

The narratives told along these routes still highlight the locations and actions of Mustafa Kemal Pasha during the Çanakkale Wars, but there are other heroes and ordinary soldiers whose actions are also celebrated, as well as times and places reserved on the tour to pray for the souls of the Çanakkale martyrs. The itineraries for these visits now include time at places such as the "symbolic cemetery" extension created in 2007 at the Çanakkale Martyrs Monument. This expanded area adjacent to the memorial has a rose garden and a forest of 59,400 red Plexiglas tombstones upon which are inscribed the names of all the martyrs from 36 different places in Turkey and former Ottoman lands who died fighting on the Turkish side. A part of the "Geography of our Martyrs" project initiated by the Ministry of Forestry and Water Affairs, the listing of the names and hometowns of these martyrs was intended to forge a closer bond with tourists who make the pilgrimage to the Gallipoli peninsula from many of these same villages, towns, and cities of Turkey and former Ottoman lands.[9] To encourage tourists to pray for the martyrs' souls, a new mosque was also erected at the edge of the symbolic cemetery grounds in 2008 and enlarged in 2018.

Diverse itineraries

Guides leading the more religiously oriented pilgrimages through the peninsula have been criticized by local tour agencies such as the Dardanelles Tourism Association for providing inaccurate information to tourists and attributing victories of the Ottoman soldiers to superstition and supernatural forces, relating apocryphal events, or celebrating the heroics of angels on the battlefields.[10] These informal developments in tourism strategies, along with official government decrees concerning appropriate visitor behavior on the peninsula, such as the 18 June 2014 ban on the consumption of alcohol in public spaces at historical places throughout the Gallipoli peninsula, have caused concern among some Turkish citizens who perceive that the religious narratives developing on the peninsula and the law restricting alcohol consumption (and the 5,000 TL fine that accompanies any violation of that law) are indications of an increasing imposition of conservative religious norms and narratives on the landscape of Gallipoli, and Turkey in general.[11]

Depending on national, political, and religious orientations, visitors to the Gallipoli peninsula follow different routes through the landscape. The itineraries of non-Turkish tourists typically highlight the iconic moments and places experienced on the peninsula by the Allied troops (primarily the ANZACs), and/or the Commonwealth and War Graves Commissions' tidy cemeteries and commemorative monuments.[12] However, most Turks visiting the peninsula will follow different routes, which can vary in direction, allocation of time per location, and the stories told. The experience of traveling to and touring the Gallipoli peninsula as a Turk can range from a day spent visiting sites mostly associated with Atatürk, hearing about his accomplishments, to a journey that includes a selection of the traditional battle locations where Ottomans won, but also adds to the itinerary more religious sites, from the aforementioned symbolic martyrs' cemetery and paradise-evoking rose gardens at the Çanakkale Martyrs Monument to several places of prayer recently restored or built anew by the Ministry of Forestry and Water Affairs or during the administrative watch of ÇATAB: Ottoman *türbes* (tombs), a *tekke* (lodge used by dervishes for the performance of the *sema* ritual of the Sufis), a *namazgah* (open-air prayer platform), the Ağadere Hospital Memorial, completed in 2015, the Akbaş Memorial cemetery constructed in 2013, or the Alçıtepe Mosque complex, completed in March 2018 with extensive space for prayer, a fountain, a social activities center, and a guest house for imams and other visiting dignitaries. These different visitor highlights and itineraries are advertised and brightly illustrated in tourist brochures and guidebooks which target specific audiences and their interests. A small selection of these guides and brochures is illustrative of the ways in which the peninsula is marketed to tourists and pilgrims with diverse interests, nationalities, native languages, and faiths.

Since 2000, among the most popular guidebooks for the peninsula in English has been Major and Mrs. Holt's *Gallipoli* in the "Battlefield Guide Series" published by Leo Cooper. The book is weighted heavily towards coverage of the Commonwealth sites but several of the major Turkish battle sites and memorials are also described. The cover image depicts the bronze casting of the "Monument of Deep Respect for Mehmetçik (the Turkish soldier)" on the Albayrak Ridge (see Chapter 1, Figure 1.6). There are various legends about the iconography of this statue, but it has become an iconic image of the narrative about the "gentleman's war" as it depicts a Turkish soldier carrying a wounded Allied soldier back to his trenches in a gesture of kindness and respect for the enemy. While the cover image showcases what is perceived to be an Australian soldier, in the introduction to the Holts' guide the authors clearly state that their intention in writing the book is not to educate Australians. Rather it is to insure that the British, among other Allied countries, are given credit for their important role and greater losses than the Australians. "If one points out to a visiting Australian, or a local Turk, that the French had almost twice as many casualties as the Australians, or that the British had almost double the losses of the other Allies put together, they are usually totally surprised and in some cases disbelieving. They are not arrogant in their belief but merely victims of a myth that has helped to forge a nation."[13]

The *ANZAC Handbook* is one of many guides prepared each year for the commemorative ceremony of 25 April by the Department of Veterans' Affairs in Canberra. The first page of the 2011 handbook shows a map of only the northern section of the peninsula, with a box around the area marked "ANZAC Battlefield Area," and clearly delineates in red lines the routes that the New Zealander and

Australian tourists are expected to follow.[14] Both the map and the commemorative program include Turkish sites and services, but the assumption is that, apart from protocol, most non-Turkish visitors will quietly bypass and not participate in the Turkish memorial services. Those coming to commemorate the ANZACs are advised: "As you walk to Chunuk Bair, please respect the fact that the Turkish Service is in progress at the Turkish 57th Regiment Memorial."[15]

Turkish brochures also show a similar lack of interest in the commemorative sites and rituals of the "other," those who have come to pay respects to the Allied dead; but among the guides produced for Turkish audiences there are distinct differences reflecting varied agendas and their itineraries. The brochure prepared by Mehmet İrdesel displays on its cover only a portrait of Atatürk when he was a soldier, with a statement in boldface at the bottom advising, in Turkish, that this is "A brochure which should be found in the hands of every Turk." Scattered throughout the brochure are several quotes attributed to Atatürk and a few pictures of the sights to be seen, among these the "place where Atatürk's famous life-saving pocket-watch was shattered."[16] The assumption here is that a visit to Gallipoli is largely a celebration of the accomplishments of the nation's founder.

Another brochure, entitled *Çanakkale Geçilmez Şehitlik Gezi Rehberi* (The Straits shall not be Breached Martyrs' Tour Guide), prepared by Salim Dağ in 2010, includes no image of Atatürk in the brochure. Mustafa Kemal's name is mentioned but his image is not presented anywhere on the pages, whereas other war heroes do make an appearance, such as Fevzi Çakmak. Of the 35 possible sites to visit on the peninsula, not one is a Commonwealth site; the corresponding map in the brochure indicates only the locations of Turkish cemeteries and commemorative places.[17] The brochure prepared by Recep Yüzüak, *Şehitlikler Gezi Rehberi* (The Martyrs' Tour Guide), displays a half-page image of the Çanakkale Martyrs' Monument but immediately below this picture is the image and invitation to stop at the tomb of an Ottoman saint, "Bayraklı Baba," or "Father Flag," a legendary figure from the fifteenth century who was alleged to have been martyred when he ingested the Ottoman flag rather than allow it to be captured by the enemy. Adjacent to Bayraklı Baba's tomb is a photograph showing the other highlights of the trip: a fifteenth century open-air prayer area *(namazgah)*, two Ottoman tombs, and a dervish lodge in the town of Gallipoli.[18] There is one small image of Atatürk which appears on the cover of another guidebook advertised in the brochure. Unfolding the map one finds a large image of a marble plaque at the site of the Turkish 57th Regiment cemetery and memorial which is inscribed with the hadith: "The martyrs requested of Allah the following: For the sake of you oh Allah send us back to the world again so that we may be martyred once more."[19]

In 2016 ÇATAB produced an official guidebook, *Gelibolu Tarihi Alanı*, or the *Gallipoli Historical Area*, which is only in Turkish and includes many of the sites on the peninsula. In this guide Turkish and foreign sites are presented loosely in the order of their location along a large loop stretching around the entire peninsula, beginning with the Akbaş Memorial and Monument, built in 2013. With 459 symbolic grave-markers, that cemetery complex after the massive 500,000 square meter complex at Ağadere, is one of the largest commemorative sites on the peninsula and is part of the "Respect for History" campaign which a national petroleum company, OPET, initiated on the Gallipoli peninsula as part of the social reponsibility component of their corporate mission[20] (Figure 6.1). *Gelibolu Tarihi Alanı* includes in its guide a total of 108 points of interest on the peninsula. Thirty-nine of these are Commonwealth or foreign sites and color-coded yellow as "Foreign Graves and Monuments"; 67 are Turkish sites and color-coded white as "Turkish Martyr memorials and monuments." Only two sites—the trenches of Kırmızısırt(53) and Arıburnu (66) are color-coded blue and indicated as points of interest for both Turks and foreigners.[21] The assumption made here in the tourist literature produced by ÇATAB is that there is little about the foreign sites or visitor experiences on the peninsula that would be of interest to Turks, and vice versa. Like the Holts' *Gallipoli*, while there is compassion for the dead of "the other" there is also the expectation that these feelings won't extend to traveling along the same paths or participating in shared rituals, prayers, or performative acts of commemoration.

Among the other publications and guides that ÇATAB has produced since 2014 is a small 20-page illustrated booklet in Turkish which is not a traditional touristic guide to the sites but a guide for how to behave properly while visiting the peninsula: the *Gelibolu Tarihi Alanı Ziyaret Adabı*.[22] The concerns ÇATAB

Figure 6.1 One of the symbolic grave markers in the Akbaş Memorial and Monument complex, built in 2013 and opened 2018.

Credit: Archive of L. Thys-Şenocak

expresses in its publications for national and local tourists are not unique to the Gallipoli battlefields; they reveal the tensions and the different expectations of and for the pilgrim and the tourist, the sacred and profane, on historic battlefields throughout the world. As Lloyd has noted, as early as December 1919 Rudyard Kipling "issued a plea to travelers to show reverence when they were on the battlefields. He called on them to remember that it is 'holy-consecrated in every part by the freely offered lives of men, and for that reason not to be overrun with levity'."[23] This small Gallipoli "book of manners" has been produced as part of the branding efforts of ÇATAB as it asserts its role as the new administrative body managing the peninsula and in response to the dramatically increasing numbers and diverse socio-economic groups that are now coming to visit Gallipoli from around the country. After recommending that visitors be aware and respectful of the fact that there was a "great loss of life of our martyrs on the lands of this important historical battlefield," it suggests that visitors stay on the marked tourist routes and be respectful of other visitors and tour groups by not speaking loudly or using devices that produce noise. Other advice given is to avoid parking cars, having picnics, lighting fires, or putting out cigarettes on the cemetery grounds. ÇATAB reminds its visitors in *Ziyaret Adabı* that the consumption of alcohol is not allowed within the borders of the Historical Area, nor is advertising, propaganda, or unlicensed filmmaking and commercial photography.

A history of architectural interventions

The one site on the peninsula that does draw a wider range of tourists, both foreign and domestic, is the Legend of Gallipoli Simulation and Information Center in Kabatepe, and it has become an important stop in the itineraries of many tourist and school groups to the peninsula since its opening in 2012.[24] Table 6.1 indicates the total number of visitors who came to the Legend of Gallipoli and the revenue and costs since its opening to the end of 2017. Visitors statistics in the five years from 2012 through 2017 indicate a steady increase in the number of visitors, with 97,220 in the opening year growing to 360,649 by the end of 2017; a total of 1,280,621 visitors have experienced the center in five years. From the official statistics provided by ÇATAB it also appears that while visitor numbers are rising, the income from the Legend of Gallipoli is not covering the costs of its operations.

In 2016 the income from ticket sales covered only 29.3 percent of the operational costs, even though this was the year with the highest visitor numbers so far. In the 2015 centenary year, only 38.2 percent of the costs of the simulation center were met by ticket sales.[25] The deficit for 2017 was 4,822,692 Turkish lira (TL), resulting in a total loss since 2012 of approximately 3,500,000 dollars. While tourism figures for Turkey have been lower overall since 2013 due to political instability and terrorist attacks, the costs of operating the center are very high and do not seem sustainable. The tour through the 11-chambered simulation center is very labor-intensive as one guide must accompany each group of 50 visitors; in 2017 alone a total of 47 people were employed just to manage the operations at

Table 6.1 Number of visitors, income, and expenses in Turkish Lira from June 2012 to December 2017 for the Legend of Gallipoli[1]

Year	*Number of visitors*	*Expenses (TL)*	*Income (TL)*
2012	97,220		553,545
2013	161,977	3,885,792[2]	829,153
2014	148,076	1,353,752	737,494
2015	222,209	2,771,775	1,058,980
2016	290,490	3,045,341	900,179
2017	360,649	5,881,625	1,058,932
Total	1,280,621	16,938,285	5,138,283

1 These are the figures provided by ÇATAB in their 2017 Activities Report available at http://catab.kulturturizm.gov.tr/Eklenti/57114,2017-yili-faaliyet-raporu-indirmek-icin-tiklayinizpdf.pdf?0

2 The Legend of Gallipoli opened on June 12th 2012 so the 2012–2013 total expense figures are combined.

the Legend of Gallipoli.[26] Part of the explanation for the wide income vs. cost disparity is that the visits to the Legend of Gallipoli are heavily subsidized by the Turkish Ministry of Culture and Tourism; in 2015, 66 percent of the visitors were exempted from paying an entry fee, 11 percent paid a reduced ticket cost, and only 23 percent paid the full cost of the entrance fee.[27]

The substantial investment in the Legend of Gallipoli and the continued subsidy of large numbers of domestic tourists, particularly school-age groups, are indications that the center was conceptualized by the Ministry of Forestry and Water Affairs as an important vehicle to promote its vision and narrative for the peninsula. Having inherited this project from its predecessors, ÇATAB continues to support the Legend of Gallipoli and hopes are high that it will generate more revenue in the future. The director of ÇATAB, İsmail Kaşdemir, announced on 14 March 2017 that he expects the number of tourists to the Gallipoli peninsula to increase in the next couple of years from 4 to 20 million.[28] While these figures are rather optimistic, it is likely that, based on past statistics, roughly 40 percent of the tourists visiting the Çanakkale region will also visit the Legend of Gallipoli.[29] For that reason, a more detailed analysis of both the architectural design and scenarios for the Legend of Gallipoli are provided below, and analyzed as part of a series of recent architectural projects intended to promote the heritage of the Gallipoli peninsula.

Law 6546 established ÇATAB on 19 June 2014 and outlines the duties and responsibilities of the directorate, which includes protecting, advertising, developing, and passing down to future generations the historic, cultural, and spiritual/moral values of the sites where the Çanakkale naval and land wars took place, all while preserving the natural fabric of the landscape.[30] With this law the status of the Gallipoli peninsula, which had been designated as a national park in 1973, changed; the management of all aspects of the peninsula was transferred from the Ministry of Forest and Water Affairs and the Gallipoli Directorate for

Nature Conservation and National Parks to ÇATAB. The latter was tasked with determining the objective, strategy, and decisions that would be needed to protect, develop, manage, and represent the site, determine the provisions for its use, protect its historic, cultural, archaeological, natural, and outstanding universal values, enable the national and international community to understand the site, and improve the social, economic, health, and transportation infrastructure of the region—all with a view to passing down the site to future generations and conceptualizing it as an open-air museum.[31]

The goal of transforming the Gallipoli peninsula into an immense open-air museum is not new. The idea was suggested, for different reasons, as early as 1918 by C.E. Bean and others who faced the prospect immediately after World War One of identifying, exhuming and reburying thousands of human remains of fallen Allied soldiers in the northern section of the peninsula. The Gallipoli Peace Park International Ideas and Design Competition of 1998, mentioned earlier,was organized by Raci Bademli and his team from Middle East Technical University in cooperation with the Turkish Ministry of Forestry. In the competition documents sent to participants it was suggested that the peninsula would be best preserved as an open- air museum, but this time one that included the entire peninsula, not just the northern battlefields where the majority of the ANZAC troops had perished. Participants in the 1998 competition were invited to propose ways to forefront a meta-narrative of peace for this unique landscape and to conceptualize it as a "Peace Park" comprising commemorative monuments, wartime remains, and nature; all this was to be done using the concept of an open-air museum.[32] This approach, according to the organizers of the competition and its jury members, would transform the Gallipoli peninsula into a protected zone where physical intervention in the landscape could be kept to a minimum.[33]

From the 121 entries submitted to that competition, from 49 countries, the first prize was awarded in 1998 to a team of architects from Norway: Lasse Brogger and Anne-Stine Reine.[34] Their plan, which selected the Ottoman fortress of Seddülbahir at Cape Helles as the location for a central Peace Park forum, proposed several footpaths around the peninsula as a way to create a reflective journey for the visitor. They conceptualized the peninsula using what Ahenk Yılmaz has described as a counter-memorializing approach.[35] Rather than constructing more commemorative monuments, Brogger and Reine suggested that visitors to Gallipoli would be better served if they experienced "a journey for the visitor which is free from all existing memorialization attitudes of different nations and periods and depends on the real memory of the landscape of war."[36] The experience of the Gallipoli landscape was envisioned as an embodied one (albeit in competition documents as a male body) which connected all visitors, regardless of nationality or faith, to their surroundings through both the power of the gaze and physical participation in the landscape. "The foot is an eye moving step by step. The eye wanders in open space. Moving on the path. Man moving, being the shuttle weaving his own history is essential to our approach to the park."[37]

The winning project was never implemented but its non-invasive approach was lauded by the jury members. The plan, according to the jury, "respects the site as it is, incorporating minimal interventions that enhance the landscape . . ."[38] Compared to other entries this was the case. Yet, if the Brogger–Reine plan had been implemented, a large section of one of the earliest and intact seventeenth century walls of the Ottoman fortress at Seddülbahir would have been destroyed to create a new entrance into the fortress' interior. While the Norwegian architects suggested that a careful restoration of the existing structure had to be undertaken, they recommended problematic restoration measures for a heritage structure: "a new entrance for visitors is made in the north-western wall" and "restored parts are executed in concrete and brick, materials not used in the existing structure."[39] Additionally, parking lots for tour buses were to be located on part of the former battlegrounds near the Cape Helles beachfront adjacent to the west wall of the fortress of Seddülbahir.

Among earlier commemorative projects on the peninsula which have purported to be non-invasive and garnered the support of competition juries are Ahmet Gülgönen's large epigraphic slabs, which were awarded first prize in the Conkbayırı Mehmetçik Park Design Competition in 1970 (Figure 6.2). [40] Introduced in the first chapter of this book, these immense epigraphic monuments narrate different stories about the victorious Ottoman soldiers, including Atatürk, who fought on this contested ridge in the northern section of the peninsula.[41] The first of the five slabs was erected at Conkbayırı and 15 more appeared at other points on the landscape, including Arıburnu, over the next decade (Figures 6.3(a) and (b)). A project initiated after the establishment of the Gallipoli peninsula as a national park in 1973, the Conkbayırı concrete slabs were registered on 14 November 1980, along with several other existing monuments on the peninsula,

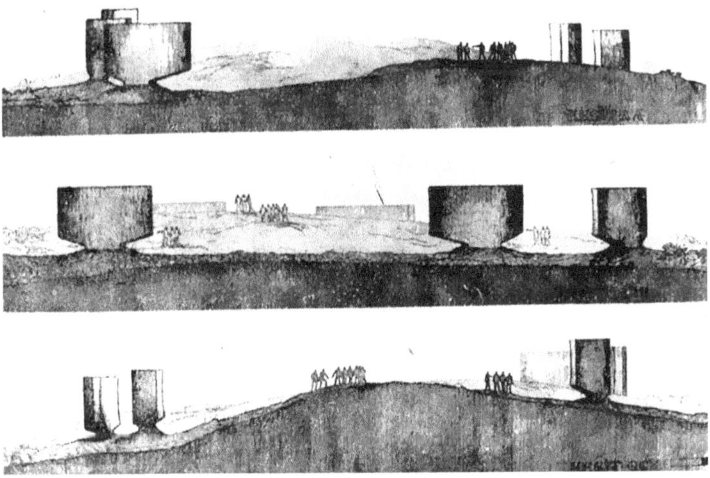

Figure 6.2 Ahmet Gülgönen's large epigraphic slabs were awarded first prize in the Conkbayırı Mehmetçik Park Design Competition in 1970.

Credit: "Conkbayırı Mehmetçik Park Anıtı," *Mimarlık*, 1970

Figures 6.3(a) and (b) Ahmet Gülgönen's epigraphic slabs at Arıburnu (a) and Conkbayırı (b) function as gigantic museum labels.

Credit: Archive of L. Thys-Şenocak

as cultural heritage, almost ten years after the completion of their initial construction. All jury members and participants in the 1970 competition were of Turkish origin; notable among the jury was Doğan Erginbaş, one of the members of the team of architects who had won the competition for the Çanakkale Martyrs Memorial, in the 1940s.[42]

Gülgönen's winning project was praised by the jurors for its non-invasive, even subtle approach to the landscape, its evocation of the Orhan inscriptions (the earliest examples of Turkic writing on monoliths in Mongolia from the sixth to the seventh century), and its reasonable cost of construction and upkeep.[43] The architect of the winning design for the Conkbayırı epigraphs "wanted just to dot the landscape with those epigraphs and not to intervene."[44] He recognized that there were human remains which would inevitably be disturbed if a large structure was erected in this area of the battlegrounds. To avoid intervention in the landscape itself, he designed a very narrow support system upon which the epigraphic slabs were to be mounted, and rejected the construction of any kind of covered pavement or visitors' path, allowing the local flora to flourish and people to find their own route to circulate and read the words on the monuments. Ahenk Yılmaz has remarked that the monuments "orient the observer to see the significant places narrated in the inscriptions from the gaps in-between each other . . . one starts to comprehend the landscape of memory of the battles by means of pre-defined strictly framed scenes. The locus of the image does not detach the visitor from the actual place completely . . . They just orient (the) individual's movement, do not guide according to a predefined path."[45]

Yılmaz's assessment of Gülgönen's design is kind. Contrary to the architect's and jury members' claims about the non-invasive quality of the project, the five monolithic slabs dramatically slice up the view of the northern shores of the Gallipoli peninsula, obscuring sections of it and making it more difficult to understand the changes in topography that were so crucial to the successes and failures of the battles fought in this section of the peninsula. Though perhaps effective as wind-breakers, these epigraphic slabs are visually invasive and do not encourage the individual to engage in any "meaning-making" process or contemplation of the surrounding landscape or the events that occurred there. Ironically, two of the jurors remarked that the epigraphic monuments served in a positive way to preserve the landscape as if it was a museum and another praised them for being human-scale.[46] The slabs do actually function like gigantic museum labels but overwhelm the visitor as both the size and tone are highly didactic and authoritative. Gülgönen's failure to anticipate and provide for visitor traffic has also led to the erosion of several sections of the site and difficulties in managing large numbers of tourists moving around the slabs. While the omission of visitor paths was due to the architect's wish not to disturb human remains, abandoning the site to the navigational discretion of large tour groups has not resulted in the environmental protection the architect intended.[47] Throughout the 1980s, even more of the large epigraphic slabs designed by Gülgönen were erected at eight other sites in the surrounding battlegrounds, fulfilling the architect's vision that these slabs would form a chain of monuments leading the visitor through different points on the battlefields of the peninsula.[48]

Gülgönen also designed the first museum and visitor center at the Kabatepe Harbor, along with a ceremonial area and parking lot.[49] This project was then expanded in 1983 by the architects Metin Hepgüler and İlhan Şahin, the winners of the architectural competition, "Çanakkale Gallipoli Peninsula Historic National Park Kabatepe Symbolic Martyrs Monument," which expanded the Kabatepe visitor center with a bigger parking lot and a 35 ton triangular metal commemorative sculpture embedded into an enlarged terrace and whose shape was intended to evoke the attack position of an enemy soldier.[50] By 2002 opinions had started to change about both Gülgönen's epigraphic slabs and the 1983 award-winning commemorative complex at Kabatepe. The former project was criticized for the poor quality of the prefabricated concrete, and the latter for both the poor choice of materials and the ineffective abstract style.[51] The museum of the Kabatepe complex was torn down in 2012 and replaced by the largest structure ever built on the Gallipoli peninsula: The Legend of Gallipoli Simulation and Information Center.

The Legend of Gallipoli Simulation and Information Center

The Legend of Gallipoli was part of a much larger project, initiated in 2005 by the Ministry of Forestry and Water Affairs to develop the touristic infrastructure of the Gallipoli peninsula. The Ministry also stated that "A long term restructuring process was instigated in the peninsula for the park to gain a 'peace identity'." [52] While never sanctioned by the ministry as an official component of the peninsula's identity, the idea of a "Peace Park" was promoted unofficially by the Ministry of Forestry and Water Affairs. During their campaign to reorganize the peninsula, the Ministry of Forest and Water Affairs tore down the original Kabatepe museum and information center and launched an architectural competition for a visitor and simulation center along with a museum on that same site.[53] The competition was met with protests from the Turkish Chamber of Engineers and Architects

(TMMOB) who announced that that it was both inadequate and illegal for several reasons, among these the fact that there was only one architect on the competition jury.[54] The proposed project was also criticized by members of the chamber for ignoring the results and the recommendations of the earlier 1997–1998 international competition initiated at Middle East Technical University by Raci Bademli, which envisioned the peninsula as a "Peace Park" with a mandate for minimal intervention in the Gallipoli peninsula.[55]

In spite of the protests by the TMMOB, the concept design competition for the Legend of Gallipoli was held, and Ebru Altay and Perin Öztürker of the Ankara-based firm of PEB Architects won first prize in 2009 for their design concept and scenario. The building opened on 7 June 2012.[56] According to the reports of the Ministry of Forest and Water Affairs, the center encompasses a total of 8,600 square meters; 2,400 square meters of that is covered space, making it the largest enclosed commemorative space ever constructed on the Gallipoli peninsula dedicated to telling the story of the Çanakkale Wars.[57] The official website of the Justice and Development Party mentions that the construction of the center cost 58.7 million Turkish lira (approximately 33 million dollars in 2012); the "edutainment" complex also contains a museum, conference facility, gift store, cafeteria, and prayer space. Its opening museum exhibition in 2012 included the redisplayed wartime objects from the earlier Kabatepe museum (Figure 6.4).[58]

According to the original plan submitted by the architects for the Legend of Gallipoli, the center was intended to be a non-invasive structure and to

Figure 6.4 The Legend of Gallipoli Simulation and Information Center.
Credit: ÇATAB

work towards the larger and frequently mentioned goal of creating an open-air museum on the peninsula. The design, wrote the architects, was informed by the juxtaposition of the antithetical concepts of war and peace and the line that separates "Peace" (characterized by belief, patriotism, determination, courage, resurrection, and a nation awakening) from "War" (characterized by occupation, death, exploitation, violence, and conflict). In their notes about the intended construction, the architects stated that no trees would be cut down, trenches destroyed, or martyrs' remains disturbed. All floors of the center were to be located underground, extending to a depth of more than 17 meters below the surface so as not to intrude on the landscape. Visitors would have the sense of entering into a commemorative space embedded deep in the earth (Figure 6.5).[59]

Above ground only two immense, curved, and opposing walls, each 9.35 meters high mark the entrance to the center. According to the architects' initial scenario, these massive walls—one black and the other white—symbolize war and peace and the physical geography of the Dardanelles, with the passage between them emulating the straits dividing the continents of Asia and Europe. The design consists of open and closed spaces of memory and commemorative platforms which emphasize the spirituality of martyrdom and monumentality (Figure 6.6). The plain white wall (originally intended to be on the left) is split at one end into three pillars representing the Turkish soldiers' faith, bravery, and love of country; its smooth surface and serene color expresses death and loneliness. The opposing wall, which is more gray than black, was originally intended to form the border of an open-air museum and commemorative space at the site "where the souls of the Turkish soldiers who became martyrs while fighting against occupation and colonialization would be honored for their civic duty" (Figure 6.7).[60]

The architectural competition for the Legend of Gallipoli was officially a competition for the conceptual design of the center. From a comparison of the plans and description for the original design submitted by PEB and the actual building that stands today in Kabatepe, it is clear that what was built in the end—both physically and conceptually—differs in significant ways from the original winning design created by the PEB architectural team. While the facility does have two underground floors, the majority of the complex is above ground, resulting in a dominating and highly visible intervention in the landscape. It is a mammoth spectacle which, due to its size and gleaming white exterior, overwhelms its surroundings and can be seen from the farthest ridges of the battlefields in the ANZAC sector.

The visitor to the Legend of Gallipoli is meant to proceed into the center of the complex via the passage between two opposing walls, thereby partaking in a performative act which separates the concepts of "peace" and "war," dividing the Allied forces on one side and the Ottomans on the other. Very little of the original concept and scenario, or what was finally realized, questions or challenges this divide, or ventures to present a more nuanced approach to the complexity of the Çanakkale Wars, or conflict in general. While intended as a visitor center for all tourists to the peninsula, the message conveyed in the design for the Legend of Gallipoli is the very divisive nature of war (Figure 6.8).

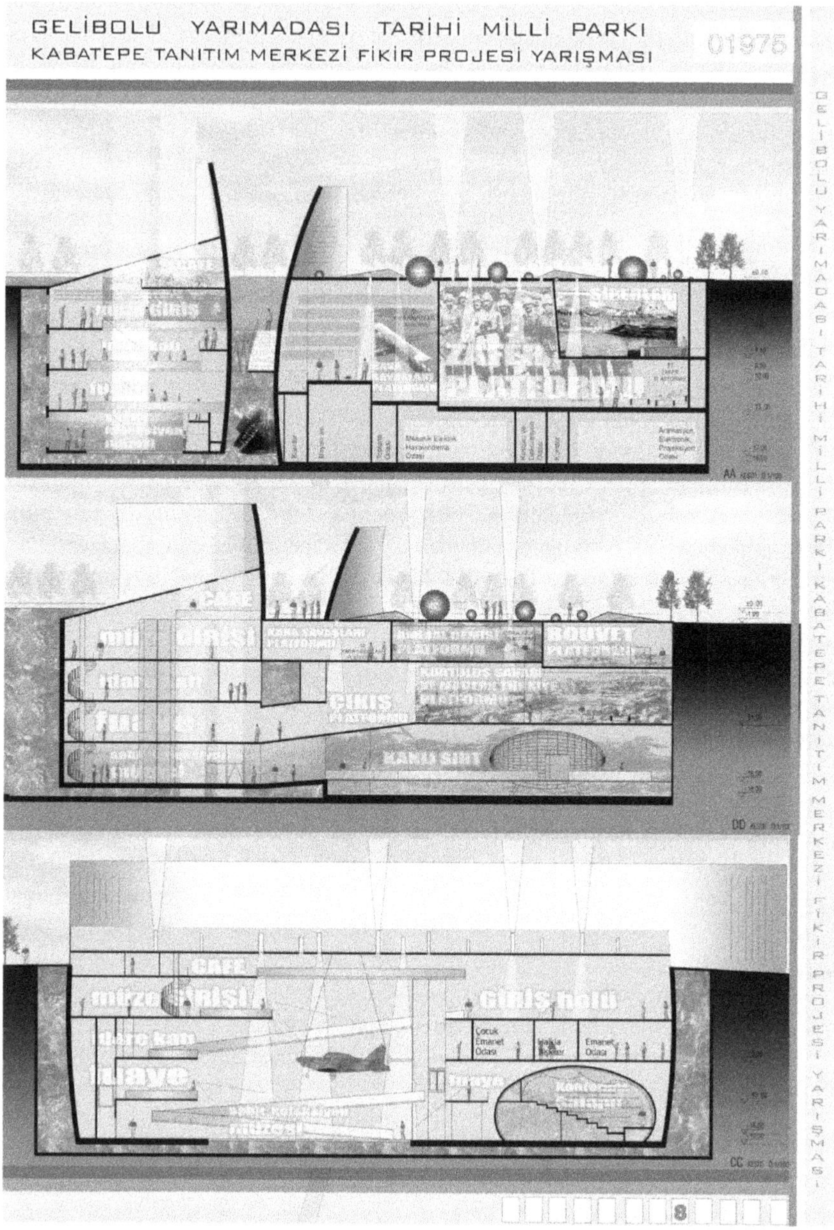

Figure 6.5 PEB Mimarlık drawings show that the center was to be located 17 meters underground.

Credit: ÇATAB/PEB Mimarlık

Figure 6.6 In the original plan the two monumental walls, one black and the other white, were intended to stand as metaphors for the opposing sides of war and peace; view from the north.

Credit: Archive of L. Thys-Şenocak

Figure 6.7 The curved white wall (left) is split into three flattened pillars representing the Turkish soldiers' faith, bravery, and love of country.

Credit: ÇATAB/PEB Mimarlık

Figure 6.8 The appendage (originally intended to be on the northern wall) resembles an Allied ship sunk during the 18 March Ottoman naval victory.

Credit: Archive of L. Thys-Şenocak

When the Legend of Gallipoli was constructed, many components of the original plan of the building were altered. The appendage on the south side of the building was originally to be on the north wall of the structure and was intended to serve as a metaphor for one of the Allied ships that was sunk during the 18 March Ottoman naval victory. The black and white color of the two walls was also reversed in the built version, as was the north wall with the three flattened pillars representing the Turkish soldiers' faith, bravery, and patriotism. This wall was rendered in a dark stone rather than white, as was originally intended by the architects. The functions of several of the interior spaces in the center were also changed; one notable alteration was that some of the conference and facility rooms were converted into a prayer space so the visitor can now, once he or she emerges from the final simulation chamber, have an opportunity and space to pray if so inspired, or continue to the museum exhibition.

The original plan for the interior of the simulation center called for eleven separate chambers, and different scenarios were prepared for each room. The architectural team proposed a high-tech multi-sensory experience which combined 3D projections, holograms, moving platforms that mimic warships floating at sea, and in some of the chambers the sound of flies buzzing and the smell of dead bodies. These latter two suggestions, while interesting in light of the recent sensorial turn in museum exhibition design, were not included in the final programs for the eleven chambers. Many other aspects of the architects' scenarios for the center departed, in some cases quite dramatically, from their original proposal. A thorough analysis of all the differences between the proposed scenarios and the final product is beyond the scope of this chapter, but a few examples are illustrative of the original narratives that were proposed by the architects and what was eventually realized by the Ministry of Forestry and Water Affairs and the committee that implemented the project.

In the first example, the scenario proposed for Chamber 3 in the original proposal was to be a simulated experience on the deck of the French warship *Bouvet*, which was targeted by the Ottomans and sank in the waters of the Dardanelles on 18 March 1915. The scenario for that chamber substituted the famous Ottoman minelayer *Nusret* for the French ship and highlighted the former commanders' devout Islamic religious beliefs, prophetic dreams, and skepticism of German military strategy as reasons for Ottoman success. The official catalog for the center mentions that visitors would now experience the rumbles of the *Nusret* rather than a French warship, and this experience would be accompanied by the "low level sounds of a prayer '*Bismillah Funda*,' animated in this chamber"[61] The story of the success of the *Nusret* is narrated in the catalog published by the Ministry of Forestry and Water Affairs for the Legend of Gallipoli.[62] In the catalog, Cevat (Çobanlı) Pasha, the commander of the area to be mined by the *Nusret*, has a dream in which he hears a voice telling him to look at the sea. "When he looked at the sea, Cevat Pasha sees (the) 'kef' and 'vav' letters (Arabic letters) embellished with flowers . . .; the divine voice says: 'Cevat, lay the 26 mines in the arsenal into the sea.' While Cevat was faltering in fear and astonishment, a silhouette appears and asks, 'What's the matter?' Cevat quickly tells the silhouette the dream he saw and the voice he heard just a little while ago. The man with an illuminated face comments, 'The heavenly light is the sign for victory. In the *abjad* calculation, the value of the letter "kef" is 20 and the letter "vav" is 6 and they make 26 in total'."[63] The story continues with the miraculous coincidence that there are exactly 26 mines left in the arsenal that were made by a Turk, but which the Germans recommended not be used. Finally, against the advice of the Germans, Cevat Pasha and the captain of the *Nusret* decide to lay the 26 mines "in the shape of two parallel lines in the strait, reciting prayers."[64] The mines are successful the following day; many of the Allied battleships, including the French *Bouvet*, were destroyed, marking the important Ottoman naval victory of 18 March 1915.

In Chamber 8, "The Battles of Dardanelles from the Dome of (the) Sky," the visitor is ushered into a domed auditorium to see a film in which Allied planes are shown bombing field hospitals on the peninsula, while the narrator and catalog point out that this action was "considered a crime of war according to the United Nations Rules . . ."[65] A fly-through of the peninsula in which the viewer feels lifted from the bonds of earth is accompanied by an explanation about how the martyrs of the Çanakkale Wars are now in heaven. Ottoman women make a rare appearance in the film as nurses tending to the wounded and dying in the field hospitals. Elsewhere they are visually absent from the stories told about the Çanakkale Wars. Only in the "Memory" chamber is the voice of a mother heard reading a letter to her son at the front, urging him to fight, to be a brave *gazi* (holy warrior), and to not fear martyrdom.

While the narratives created for the simulation center mention Mustafa Kemal Pasha's (Atatürk's) role in the Gallipoli battles, they mark a sharp contrast to the earlier more secularist and Atatürk-centered explanations about the war that were produced during the decades prior to the advent of the Justice and Development Party. In the Legend of Gallipoli scenarios Atatürk appears as a hologram in Chamber 6, but both Mustafa Şevki Doğan and Yahya Çavuş are equally celebrated as war heroes in the next chamber. Atatürk is not showcased as the sole hero of the battles at Gallipoli but as one among others who helped to defeat the Allied forces on the peninsula. As mentioned earlier, this revisionist interpretation, which credits Atatürk with success at Gallipoli, along with other military leaders of the Ottoman troops, is supported by many Ottoman military historians, but it represents a major departure from the earlier state-generated and more secularist storyline of Gallipoli, and the centrality in that narrative of the heroism of the Turkish nation's founder.[66]

This shift in the narrative is again evidenced in the selection of quotations for the catalog and centenary museum exhibition panels at the Legend of Gallipoli that are attributed to Atatürk. In the majority of these it is Mustafa Kemal Pasha and his soldiers' piety which is emphasized. The commander urges his troops: "No fear! Those who know how to read have Qurans in their hands (and are) getting ready to go to paradise. The ones who do not know how to read are marching reciting 'Kalima Shahadah' " (Islamic declaration of belief, which means 'God is great and unique, and Muhammed is the messenger of God')." Another version of this quote also appears on one of the exhibition panels in the museum at the Legend of Gallipoli with the heading "From the Battlefront." In this panel Atatürk's description of the battle and the unwavering faith of the Ottoman soldier is described: "They saw those who got killed, and they knew they would be killed in the next three minutes, yet they showed no minor sign of anxiety; not shaken up a bit! Those who could read the Quran did so to go to heaven. Those who did not walked to death as they recited the prayer of the kalima shahadah . . . This is an astounding example to be applauded, showing the strong spirit of the Turkish soldier."[67]

The final chamber of the center was intended by the PEB architects to highlight the War of Independence and the transformation of Turkey into a modern republic through the legacy of Atatürk; the drawings in the project archives display an image of Atatürk leaving the parliamentary building in Ankara as a way to highlight his role in making Turkey modern. While some images of Atatürk do appear in the film created for this chamber, the majority of the scenes that were eventually chosen for this chamber highlight recent achievements of Turkey and its present government with displays of neat formations of fighter planes, spotlessly clean factories, and studious students in research laboratories. Among the messages of this last chamber is that Atatürk and his legacy do not play the major role in the story of the nation's continued success and present day achievements. As one scholar has noted, "Gazi Mustafa Kemal is given thanks but the visual images are of then Prime Minister Erdoğan and former President Abdullah Gül leading Turkey to higher levels of success."[68]

Along with the simulation chambers, the first museum exhibition at the Legend of Gallipoli also opened on 7 June 2012 and continued until 23 April 2015 when the galleries were renewed for the centenary. The 2012 exhibition was an installation on two floors of the simulation center of various wartime objects and ordnance, including grenades, gunpowder, and bullets. These, and even human remains of unidentified soldiers from World War One, were displayed with very basic labels in glass cabinets (Figure 6.9). Objects from the former Kabatepe museum collection and the private collection of the Saim Mutlu museum were also re-exhibited in the glass cabinets, most placed on display mounts of deep red velvet. In the assortment of cases at this exhibition, Atatürk was represented only by an aniconic display of his uniform and boots (Figure 6.10). Like the narration of events in the simulation center script where his role in the Gallipoli battles was shared with others, Atatürk's personal effects were also presented in an exhibition

Figure 6.9 The 2012 exhibition including grenades, gunpowder, other ordnance, and human remains.

Credit: Archive of L. Thys-Şenocak

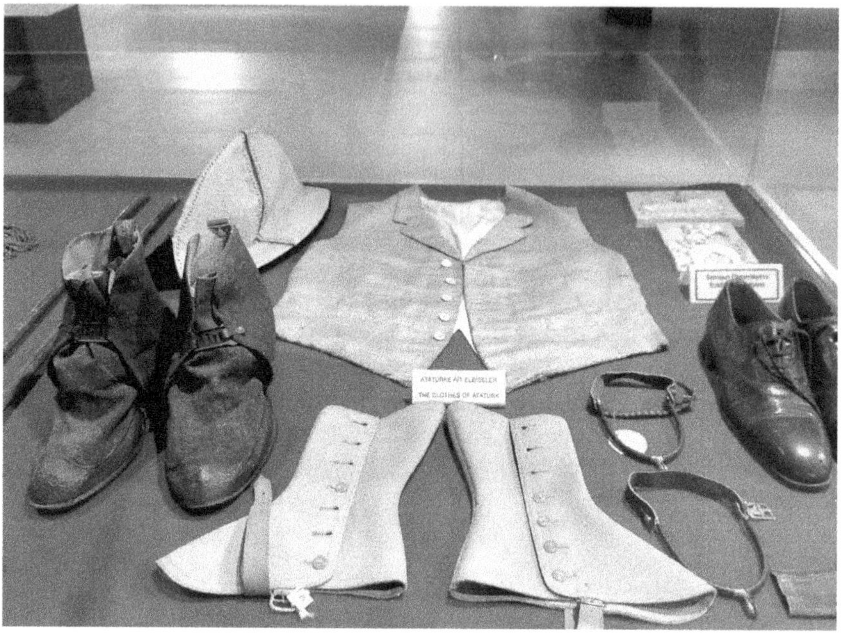

Figure 6.10 Atatürk was represented by a display of his clothing, along with those of another officer.

Credit: Archive of L. Thys-Şenocak

case with others, including the uniform of Eşref Efendi, another notable Ottoman officer who fought at Gallipoli.[69]

The Gallipoli peninsula through the lens of gender

As mentioned in the introduction to this book, due to the events of World War One that occurred on the Gallipoli peninsula the discourse which has shaped this primary identity of the Gallipoli landscape today has been highly androcentric. More archaeological research and the discovery of different layers of the past may help to bring a more nuanced and gendered analysis of the Gallipoli landscape and eventually reshape some of the interpretive strategies used by museums on the peninsula. The role played by women in the Çanakkale Wars, and any mention of gender for that matter, was entirely absent from the first museum exhibition that opened at the Legend of Gallipoli in 2012. When the permanent exhibition galleries were redesigned in 2015 for the centenary some effort was made to rectify this oversight by including in one of the cases a plastic male mannequin who was dressed as a female nurse. This figure was then surrounded by glass medicine bottles associated with her profession. This was the only acknowledgment in the museum section of the Legend of Gallipoli that women had a place in the Çanakkale Wars narratives and represents a lost opportunity, as there has been a significant amount of research about Ottoman women and their contributions to the war efforts in both World War One and the War of Independence.[70]

Most of the centenary exhibitions of World War One in Europe and Australia highlighted the roles of women in war, and some went on to explore more controversial issues related to gender roles and war, such as prostitution, rape, and transgender soldiers. Along with the narratives developed for the different chambers in the Legend of Gallipoli visitors to the museum exhibitions of 2012 and 2015 left with the impression that women were of little importance in the Çanakkale Wars. Any roles they played were subservient and traditional: obedient and devoted mothers who willingly gave up their sons to the service of the state; wives and daughters who mourned the dead, or very rarely worked as nurses who tended to the wounded.[71] David Lloyd has remarked on how an imagined divide between the home front and the battle front in Europe created the perception of a division between men and women's roles both during World War One and after it.[72] While this divide has been addressed elsewhere in several international centenary exhibitions, in the museum exhibitions at the Legend of Gallipoli the gender divide was reinforced and remains unquestioned. There is little on the peninsula that challenges the traditional views about women, gender, and war, either inside the Legend of Gallipoli or outside of its doors.

The majority of the World War One narratives that have been constructed about the landscape of the peninsula have served to gender this geography as male. This is largely due to the dominant place that World War One has had in shaping the primary identity of the peninsula. Some attempts by scholars working on Gallipoli have been made to reread, or at least to bring critical perspective to the interpretations of some of the tangible and intangible World War One heritage of the peninsula by looking at it through the lens of gender. The potential for different voices to enrich and diversify the hegemonic male discourse that now shapes the visitor experience of the Gallipoli landscape and its commemorative scenarios can be illustrated in a few examples. One author has analyzed the iconography of the sculpture "Monument of Deep Respect for Mehmetçik (the Turkish soldier)," mentioned earlier as the cover image on Mrs. and Major Holt's *Gallipoli* guidebook (Figure 1.6). The Turkish statue has taken on iconic significance as testimony to the "brothers-in-arms" camaraderie between the Turks and Allied soldiers. Adam Jones has a different, more critical reading: "The Turkish Mehmet (ordinary soldier) cradles the supine, disempowered, emasculated British captain in a manner evocative of Michelangelo's Pieta, with the crucified Christ sprawled in his mother's arms . . . For the Turkish masses who throng to these battlefields in school groups and tour groups, the sculpted 'Mehmet' is a paragon of hegemonic and nationally imbued masculinity."[73] Ahenk Yılmaz has also commented on the frequent choice of the obelisk for both Turkish and Commonwealth monuments and notes that the obelisk, in addition to being a symbol of a sunbeam pointing to the heavens like a shaft, due to its linearity and verticality of its image is also interpreted as a phallic symbol.[74] In contrast, Atatürk's comforting words to the Australian pilgrims who came to visit their lost loved ones on the Gallipoli peninsula in 1934 seem to recast the landscape of the peninsula as a nurturing and forgiving mother: "You, the mothers who sent their sons from far-away countries, wipe away your tears. Your sons are now lying in our bosom and are in peace." Gallipoli here is portrayed as a tender and welcoming body, one that is implicitly female rather than male.

The discourse used in the UNESCO World Heritage application which also gendered the peninsula as a site of a "gentleman's war" has ignored the much longer and older associations of the Gallipoli peninsula as the land bordering the Hellespont, and the associations of the region with its eponymous Helle who met her end in the Dardanelles (Figure 1.1). Apollodorus tells us about her death, and how Helle "slipped into the deep and was drowned, and the sea was called Hellespont after her."[75] Centuries later, one of the largest examples of tangible heritage on the Gallipoli peninsula was constructed by Hadice Turhan Sultan, the mother of Ottoman Sultan Mehmed IV in the mid-seventeenth century.[76] Ottoman chroniclers such as Abdi Abdurrahman Pasha celebrated the beneficence and charity of this powerful queen mother and praised her support of the empire and her architectural endeavours.[77]

> What an endeavour this is, to fill the sea on either side;
> She built the foundations for a new strait which has no equals.[78]

Heavily bombarded during World War One, the fortress of Seddülbahir stands today across from the Cape Helles Memorial—restored but partially a ruin—serving as one of the few tangible examples of the pre-war heritage on the peninsula which suffered through the violence and destruction of World War One. According to those who live today among the ruins of Turhan Sultan's fortress, this building also stands on the landscape of the Gallipoli peninsula as testimony of a woman who sacrificed her personal wealth and contributed to the defense of the Ottoman Empire and its survival. Many of the women residing in Seddülbahir who were interviewed for an oral history project conducted there in 1999 could recall easily the name of the sultan's mother and identified with the sacrifices and contributions she made as a mother devoted to the Ottoman state. As Zehra Şener explained, "There once was a Hatice Sultan, Hadice Hanım; she built the fortress with (the money from) her gold slipper . . . I heard it from my father . . . the story has always been told like this."[79]

A legacy of contested agendas

The extent to which the victory of the Ottomans over the Allied forces in the Çanakkale Wars should be attributed more or less to Atatürk and/or religious faith are hotly contested issues in Turkey, as is the degree to which the state and these different narratives recognize the role played by non-Muslim Ottoman soldiers, who also fought and died in the war. These debates will no doubt continue into Turkey's centenary in 2023 as the country struggles with the role of religion, nationalism, and minorities in its governance and the balance of these in the larger political landscape. During the decade leading up to the centenary of the Gallipoli campaign there was a significant increase in the production of histories written about Gallipoli in World War One. How these events have been remembered and reshaped for specific political agendas in various media, from poetry to official textbooks, historical novels, and films, has also received more attention since the centenary of that war. Certainly there has been an marked increased in more conservative religious practices on the Gallipoli peninsula since 2005, but it is important to note that the process of rewriting and reinterpreting the events that occurred on the Gallipoli peninsula during World War One is not a new phenomenon, nor has it been limited to one particular political party or faction of Turkish society. Some historical perspective on this is useful.

Mehmet Âkif Ersoy's epic poem "To the Martyrs of Çanakkale," published in 1924 (and introduced earlier in this book in Chapter 2), was written using the same rhyme scheme as the famed *Masnavi* written in the thirteenth century by the poet Rumi. It emphasized the important role of the Islamic faith for those Ottoman Muslims who fought on the Gallipoli peninsula in the lines, "Martyr son of martyr ask me not for a grave; the Prophet open-armed awaits his warrior, his warrior brave."[80] Nihal Atsız, a militant nationalist, was particularly critical of the non-Turkish minorities who fought on the Gallipoli peninsula in the Ottoman army; he conflated the struggles and the sites of key battles during World War One and the Turkish War of Independence while also promoting a pilgrimage to the battlefields of Gallipoli as a form of the Haj in his 1932 *Çanakkale'ye Yürüyüş*, the "March to Çanakkale":

> Oh, Turkish Youth! After leaving Mohammed the Arab's grave to those traitor Arabs who slit the throats of Turkish captives in return for British gold, is your Ka'aba not Çanakkale, Sakarya and Dumlupınar? Do you want to travel to your Ka'aba on a comfortable boat, listening to a jazz band; or would you rather travel there on foot, walking the precipitous roads and suffering as those who ran to protect the Turkish homeland in Çanakkale once did?[81]

The events of World War One on the Gallipoli peninsula have been used repeatedly in the past century by poets, prime ministers, and presidents to advocate a wide range of agendas. Şafak Horzum notes how Bülent Ecevit, four-time Prime Minister of Turkey and a member of the Republican People's Party (the oldest political party in Turkey, founded by Atatürk in 1918), wrote the poem "Gallipoli: A Postwar Epic" in 1988; it was read at the United Nations by the Prime Minister and struck a more conciliatory and even nostalgic note, particularly in the following lines: [82]

> a paradise on earth, Gallipoli
> is a burial under the ground
> those who lost their lives in fighting
> lie there mingled in friendly compound
> lying side by side
> as friends in each other's arms
> they may sleep in comfort and peace
> in the land for which they died.[83]

The Turkish General Kenan Evren, who came to power in Turkey through a military coup in 1980 and served as the seventh president of the country until 1989, hailed the victories in Gallipoli as evidence of how the military had again come to the rescue of the nation, and he also pointed out that the duty of the Turkish soldier, and the definition of the true martyr, was someone both patriotic and with faith, a person who had a "religious duty to love one's country and sacrifice one's life for its unity . . ..″[84]

Australian Prime Minister John Howard's controversial move to proclaim in 2003 the ANZAC graves of Gallipoli as the first nomination for inclusion in the National Heritage List of Australia was discussed in the first chapter of this book to illustrate how the use of Gallipoli for political agendas is not only a Turkish phenomenon. Over a decade later, in the centenary speech given at the 2015 Dawn Service on ANZAC Cove, the Australian Prime Minister Tony Abbott spoke of the ANZACs with gratitude but emphasized the patriotic duties of the citizens and soldiers of Australia in the region:

> Those serving on peacekeeping and relief missions have likewise kept faith with the original Anzacs. Even now, our armed forces are serving in the Middle East and elsewhere, defending the values that we hold dear . . . So much has changed in one hundred years but not the things that really matter.[85]

The power of bringing together patriotism and faith was communicated clearly in the controversial commercial "Dua" or "The Prayer" that was broadcast on 20 April 2015 by the Turkish presidency for the 100th anniversary of the Gallipoli battles. The video opens with soldiers who have flocked to the Gallipoli battlefield from many different provinces of the Ottoman Empire to state their name and willingness to fight in a roll call which is interspersed with images of their yet-to-be born grandchildren now living in the different regions of present day Turkey. A voice asks us in a whisper, "Have you ever heard the voice of our martyrs?" and then President Erdoğan's voice begins to recite the lines written by the nationalist poet Arif Nihat Asya in 1967, which begin, "We all have a hoarse voice . . . do not leave our minarets with no calls to prayer, my God!" and includes lines such as "Do not leave these masses, who look for a hero, with no hero, my God, Let us know how to resist the foe, do not leave us lifeless, my God." The video continues with images of Erdoğan praying at the Çanakkale Martyrs Memorial and laying a commemorative wreath at the grave of one of the Ottoman soldiers in the symbolic cemetery.[86]

Jenny Macleod has remarked that during the centenary events of 2015, the Turkish government used the occasion to deflect international attention and criticism about Turkey's official stance concerning the controversial date of 24 April 1915, the day many European countries mark as the beginning of an Armenian genocide during World War One.[87] Macleod notes that "Turkey deliberately aimed to overshadow genocide commemorations through the timing of an expanded program of Gallipoli commemorations."[88] The 18th of March is the traditional day when Turkey has commemorated the Ottoman naval victory at Gallipoli, and since 2002 this date has been officially proclaimed by the Turkish parliament as a national commemoration day on which the martyrs and veterans of all the wars in which Turkey fought are remembered. The Turkish government's move in 2015 to schedule 24 April for World War One commemorative activities was therefore considered by some to be an overtly political tactic. In response to an official invitation extended by the President of Turkey to the President of Armenia Serzh Sargsyan to attend commemorative events in Çanakkale on 24 April 2015, Sargsyan sent an open letter stating:

> Alas, Turkey continues its traditional policy of denialism. Year by year, 'improving' its tools of history distortion, this time Turkey marks the anniversary of the Battle of Gallipoli on April 24 for the first time, while it began on March 18, 1915 and lasted till late January, 1916. Furthermore, allies' land campaign—Gallipoli land battle—took place on April 25, 1915. What purpose does it serve if not a simple-minded goal to distract the attention of the international community from the events dedicated to the centennial of the Armenian Genocide?[89]

Of far more concern to the Turkish government since the 2015 centenary of the Çanakkale Wars has been the cultivation of a narrative which compares the Ottoman heroes and martyrs who fought in the World War One battles against the foreign invaders of the Gallipoli peninsula with those who defended Turkey against in the unsuccessful coup of 15 July 2016.[90] The 18th of March is the date still commemorated in Turkey as the official veterans' commemoration date, but that date, and the success that Turks achieved during World War One in driving out their enemies, has increasingly been aligned with the most recent events of 15 July 2016. In several speeches given by President Erdoğan since the failed coup, parallels with the Çanakkale Wars have been drawn. In one speech, the president promised to build a commemorative monument similar to the Çanakkale Martyrs Memorial on the Asian side of the Bosphorus Bridge now named the 15 July Bridge, in honor of those who lost their lives defending Turkey on that evening of the attempted coup.[91] In another speech on 18 March 2017, the descendants of the foreign enemies in the World War One battles of Çanakkale, or those opposing the introduction of the presidential rather than parliamentary system in Turkey, were targeted: "They haven't yet forgotten the defeat inflicted on them a century ago in Çanakkale by the Turks, whom they called 'sick man' and extended their condolences to one century ago. They are well aware of the fact that the presidential system of government will pave the way for new Çanakkale victories in every realm from politics to economy and diplomacy to investments."[92]

On 13 July 2017 President Erdoğan delivered a speech entitled "July 15 Is Our New Çanakkale" at the Commemoration Event for July 15 Martyrs at Beştepe People's Culture and Congress Center.[93] Standing in front of a large poster bearing the text "*15 Temmuz Destanı*" ("The Legend of 15 July"), the President left no doubt that he intended the defeat of the enemies of the Turkish nation during the previous year's attempted coup to be considered equal in historic importance, and to have the same impact on the country's future, as the Ottoman victory in the Çanakkale Wars a century earlier:

> States and nations have critical turning points in their histories that shape their future. The 15th of July is such a date for the Republic of Turkey . . . Our goals for 2023 are the name of the means that will widen our horizon, broaden our range, give strength to our knees and arms, refresh our breath and establish a bond between our past and future. After the sufferings endured following the defeat in the Balkans, our ancestors won one of the greatest victories not only in our own history but in the global history of warfare. Did we have the technology they had? No. But we had our faith. With the courage and self-confidence Çanakkale gave us, we carried our War of Independence to victory. July 15 is our new Çanakkale, Dumlupınar and Sakarya.[94]

Conclusion

While the 2015 centenary of the Gallipoli campaign was an important milestone, the planning framework for ÇATAB and the current Turkish government since 2014 has been concerned with gearing up to fulfill several of its goals in time for the centenary celebrations of the Turkish Republic, which will take place on 29 October 2023.[95] Among the key projects in the Çanakkale region is the construction of the longest suspension bridge in the world, the 1915 Çanakkale Bridge, which will connect the European and Asian shores of the Dardanelles, and is slated to open in 2022. The groundbreaking ceremony for the 3.7 km bridge took place on 18 March 2015, the national Martyrs Memorial Day. This ambitious project is just one of many being planned by the central government, together with ÇATAB, for Turkey's centenary and as part of the 2016–2020 strategic development policies for the Çanakkale region. Çanakkale and Gallipoli will no doubt be an important component of the 2023 vision for Turkey; President Erdoğan stressed again, in his speech for the 18 March 2016 National Martyrs Day, that Çanakkale and the Gallipoli battlefields will play a role in the Turkish centenary of 2023:

> All methods of war of the era were employed at sea, on land and in the air to the greatest extent during the Battle of Çanakkale but it was shown to the entire world that the true power is not technology but faith . . . History books tell us about the momentous consequences of the Battle of Çanakkale. We will live, Allah willing, and see what kind of historic changes the current developments in our region will bring about . . . In this process, it is our responsibility to hang on to and persevere in our goals for 2023.[96]

Between 2014 and 2016, ÇATAB pursued and developed its goals in several workshops devoted to strategic planning for the peninsula. While it is still too early to evaluate the effectiveness of this new administrative organization, a brief examination of some aspects of the site management policies proposed for the peninsula, and the 2016–2020 strategic planning documents, can help to illuminate some of its goals for the future of Gallipoli.[97] Based on the official publications (e.g., tourist guides, workshop reports, planning documents for site management, annual and mid-year reports) that it has generated since its inception in 2014, ÇATAB is clearly trying to distance itself and differentiate its goals and management practices from those of its predecessor, the Ministry of Forestry and Water Affairs. Some of these changes are nuanced, but important. In its 2017 mid-year report, ÇATAB noted that there are plans to change the scenario promoted in the Legend of Gallipoli and correct erroneous information in the scenario and exhibition.[98] The Gallipoli Historic Region Guide (*Gelibolu Tarihi Alanı Rehberi*), the aforementioned guidebook published in 2016 under the auspices of ÇATAB and the Turkish Ministry of Culture and Tourism, also makes a point of noting that particular sites on the peninsula, such as the symbolic cemeteries, were developed or restored by the Ministry of Forestry and Water Affairs without sufficient archival research or scientific documentation.[99]

Less subtle are the goals expressed in the 2016–2020 strategic planning document and the 1/25,000 scale Historic Region Planning Document (*Çanakkale Savaşları Gelibolu Tarihi Alanı 1/25,000 Ölçekli Tarihi Alan Planı*) commissioned by ÇATAB for the peninsula. Both documents are extensive and reiterate the importance of now well-established and internationally accepted management practices for heritage sites, such as sustainability, participatory decision-making processes, environmentally friendly tourism, and more organic farming methods. According to these plans, laws will be passed to increase ÇATAB's jurisdiction to protect underwater heritage within 500 meters of the shore, particularly to protect the many shipwrecks from World War One and earlier eras in the Dardanelles and its surrounding waters. More ferryboat transport lines will be developed to connect the peninsula and the historic sites of the Asian side of the Çanakkale region, creating a touristic ring route between the Ottoman and World War One sites on the opposite shore of the Dardanelles, such as Kumkale and ancient sites such as Troy. The peninsula will become a model of accessibility for visitors with disabilities. Specific projects for the peninsula are proposed, such as the development of a new cultural route dedicated to the Ottoman fortifications on the peninsula and the possible reconstruction of the Decauville narrow-gauge railway that was used to move ordnance and other supplies around the peninsula during World War One.

The list of projects and goals for 2023 is extensive and ambitious, but some are particularly noteworthy as they depart from the planning principles and goals of the earlier administrative body (the Ministry of the Forestry and Water Affairs), and some actions are a response to the wave of faith tourism that has been increasing since 2005. For example, there are to be no more symbolic cemeteries constructed in the areas administered by ÇATAB; the graves of actual martyrs, once verified through archival and archaeological research, will be incorporated into the larger organization framework: an open-air museum.[100]

It is not clear yet how ÇATAB defines an open-air museum or what lessons it has learned from inheriting the Ağadere Hospital Memorial complex, and Akbaş Memorial and Monument project from the former Ministry of Forestry and Water

Affairs. At both Ağadere and Akbaş large tracts of land were transformed into immense commemorative complexes. At Ağadere, the architect Oktan Nalbantoğlu and his team won the 2012 competition that was held by the Ministry of Forestry and Water Affairs for this site where over 2500 Ottoman soldiers were buried. The project, completed in 2015 involved the creation of several kilometers of paths throughout 500,000 square meters of the Ağadere valley one kilometer north of Kilitbahir, along with the construction of 13 memorial structures, a massive exhibition hall, and a mosque, the latter two made of pre-cast concrete. For the Akbaş Memorial and Monument Project symbolic tombstones for 459 martyrs were all built adjacent to the original Ottoman burial grounds.[101] While the Ağadere complex, completed in 2015, has not yet opened and is an abandoned mass of concrete and weeds, the Akbaş Memorial opened on 14 January 2018 and includes a parking lot, two separate *mescids* for men and women, *abdesthanes* to wash prior to prayer, toilets, a tea stand, and five booths for selling various tourist items. A commemorative wall stretching 35 meters across the entire complex is inscribed with the names of 754 Ottoman martyrs from the war and flanked by two sculptures of soldiers.

The petroleum company OPET financed the construction of the Akbaş memorial as part of their "Respect for History" corporate social responsibility campaign for the Gallipoli peninsula, initiated in 2006. The OPET website pronounces that through its generosity, in addition to the 7,500 square meters of paving stone used to cover the site, they contributed 3000 stones to build a massive masonry wall, paid for 7500 days of labor, and used 25 machines in the construction. Another construction, the Alçıtepe Mosque complex in the village once called Krithea or Kirte eight kilometers to the north of Cape Helles, was also financed by OPET and Turkish Air Lines (THY) and opened for prayer on 18 March 2018.[102] While still not challenging the 500,000 square meters of land used for the Ağadere project; the Alçıtepe mosque complex encompasses twice the surface area of the Akbaş or the Legend of Gallipoli projects; the mosque complex was allocated 15,000 square meters in the center of the village which was the location of three major battles of the Çanakkale Wars in April, May, and June 1915 (Figure 6.11a).

Several studies have praised the social responsibility projects undertaken by OPET's Respect for History campaign, citing the fact that the corporate contributions are sustainable and have been met with overwhelming approval by residents of the peninsula.[103] One study published in 2018 noted that the situation on the Gallipoli peninsula had vastly improved when the "lack of attention to the region was changed by the increasing state and private-sector interest in recent years." This same study also praised OPET for its Respect for History project because it had become a case study used at Harvard to teach corporate cultural responsibility and: "This project, eagerly brought to life in a short time, contains a rehabilitation process and trainings that do not ruin the natural texture of the region."[104] A primarily Greek village since the fifteenth century, Alçıtepe, known for its barley (*krithari*) in the Antique world, was also one of the oldest continuously populated villages on the Gallipoli peninsula. In spite of its historic importance no scientific archaeological work was conducted prior to the construction of the Alçıtepe mosque complex, so all archaeological evidence from World War One, the Ottoman era, and earlier eras of Alçıtepe's past has been lost, without any scientific documentation.[105]

The SWOT (Strengths, Weaknesses, Opportunities, and Threats) analysis conducted by ÇATAB in 2016 for their strategic plan assesses the current and future threats that are perceived by the directorate. Among the perceived list of threats are the problems of managing too many visitors, and the insufficient capacity of the Gallipoli peninsula to accommodate these crowds. Another threat identified is the possibility of visitors being given incorrect information about the peninsula that is based on superstition and apocryphal stories, the abuse and exaggeration of nationalist sentiment, the distancing of the visitor from feelings about peace, and the promotion of exaggerated and incorrect information.[106] To manage the apocryphal tales told about the war by private tour operators, starting in 2018, guides for the peninsula will have to be licensed by ÇATAB and attend a training course, whose curriculum will be dictated by ÇATAB.[107] One threat not mentioned in the SWOT analysis is the problem that private financing of different heritage and commemorative projects could bring in terms of the branding of the peninsula. With the increasing amount of corporate signage, who is responsible for the management of the Gallipoli peninsula and its identity is becoming more difficult to determine.

The winner of the most recent architectural competition, the "Gallipoli Historic Region New Martyrs Concept Design Project Competition" (*Gelibolu Tarihi Alanı—Yeni Şehitlik Tasarımları Fikir Projesi Yarışması*) was announced by ÇATAB on 25 July 2017 and awarded on 9 January 2018. The architectural team led by Sema Uygur proposed a concept project which is intended to link 15 sites on the peninsula which are significant for memorable events that occurred during World War One and/or burial sites of Ottoman martyrs. The "Path of Respect" proposed by the team stretches along the north–south axis of the peninsula for approximately 35 kilometers and is interspersed with sections of a low travertine "Memory Wall" (Figure 6.11b) upon which names of Ottoman martyrs will be inscribed.[108] The team was awarded first place by the jury for its non-invasive approach they took to the landscape and the team's use of the meta-concept "necrogeography."[109] This is a term that came into use in the 1960s to describe mortuary landscapes and their spatial as well as cultural meanings; according to Matero and Peters, the preservation of this type of landscape requires "an understanding of the origins of each site, as well as how changes in customs and attitudes over time have affected a site's use, form, meaning, and preservation."[110] The jury in this most recent architectural competition on Gallipoli also noted that by using stones from around Anatolia, the winning proposal successfully united the landscape of the peninsula with the many different geographies from which the different martyrs of the war hailed. Praised for its lack of monumentality and ornamentation, the "Memory Wall" and the "Path of Respect" draw upon other iconic architectural projects and paths of commemoration and resurrection both within the Gallipoli Peninsula at Ağadere, and outside of Turkey, most notably the 1982 Vietnam Veterans Memorial by Maya Lin in Washington D.C., and the refurbished High

Figures 6.11(a) and (b) The Alçıtepe Mosque complex opened in March 2018 (a) the proposed "Memory Wall" (b) was the winning entry of the Gallipoli Historic Region New Martyrs Concept Design Project Competition, 2017.

Credit: Archive of L. Thys-Şenocak (6.11a) and ÇATAB (6.11b)

Line, a section of railroad on the west side of New York City which was brought back to life as an elevated walkway in 2009 by the architectural firm of James Corner Field Operation.[111] Appearing on paper as less invasive than the epigraphic slabs designed in 1970 by Ahmet Gülgönen for the Conkbayırı Mehmetçik Park Design Competition, the "Memory Wall" and "Path of Respect" project, if and when implemented, like the Ağadere, Akbaş and Alçıtepe commemorative projects, risks becoming yet another example of a major intervention in the landscape. Like so many of its predecessors, this project too has been lauded by its architects, jury members, state and corporate sponsors, for being non-invasive, for furthering the goal of an "open-air" museum, and for working to respect and integrate seamlessly with the landscape. But in essence it too is another wall, another structure imposed upon and invading the Gallipoli landscape.

David Lloyd has pointed out that in contrast to the post-war landscapes of Europe—landscapes like Ypres, which were vigorously rebuilt after the destruction of World War One—"Gallipoli was not reconstructed after the war because there were no villages to rebuild . . . The untouched landscape enabled many travelers to feel closer to the dead."[112] As this newest phase in the life of the Gallipoli peninsula unfolds, those who manage the peninsula as stewards of such a rich and layered cultural landscape, and the architects who compete to make their mark here, must realize the vast potential this region has to tell us about a diverse array of ancient and recent pasts, and heritage narratives: about the dead, and about the living. The many stakeholders who are so passionate about the peninsula must also realize that even well-intentioned projects aimed at commemorating Gallipoli's past—its martyrs, its tangible, intangible, and living heritage—have resulted in repeated interventions in the landscape during the last century. Most of this has been done without proper archaeological documentation, and vital information about the region's past has been irretrievably lost. For the layers of the Gallipoli landscape in which the events of the Çanakkale Wars exist, it is essential that the very earth in which these martyrs are buried is regarded with care and respect. As Price has noted, archaeologists "are themselves part of various communities for whom the landscape and remains are sacred." As the poets and politicians remind us, these martyrs have become a part of that earth now. Any intervention in this landscape, for any purpose should begin with that premise.[113]

Among the most important heritage guidelines to have evolved in the past decades is the Burra Charter. First drafted as a charter for Australia in 1979, it has been revised several times and has had an impact far beyond its national borders as it contains many recommendations for how to strike a balance between the use and preservation of heritage sites. It advises, for example, that if we are to use a place of heritage we must be certain to respect the cultural significance of that place. "Such a use involves no, or minimal, impact on cultural significance."[114] Another heritage concept codified in Canada is the "spirit of place"; like the Burra charter, it suggests strategies for preserving the tangible and intangible characteristics of a site in the face of threats to spiritual significance. Article 4 of the Quebec Declaration of the Preservation of the Spirit of Place states:

> We recommend that governmental and non-governmental agencies and local and national heritage organizations develop long term strategic plans to prevent the degradation of the spirit of place and its environment. The inhabitants and local authorities should also be made aware of the safeguarding of the spirit of place so that they are better prepared to deal with the threats of a changing world.[115]

There is no lack of heritage instruments to employ in the stewardship of the Gallipoli peninsula. If there is a real need on the Gallipoli peninsula for more places of commemoration, more spaces for prayer, more parking lots, wider roads, or edu-tainment centers for tourists, then the impact on the cultural significance of this landscape needs to be clearly evaluated for what is lost and what is gained. As ÇATAB prepares the Gallipoli peninsula for nomination to the World Heritage List and prepares for Turkey's centenary, those who are managing the peninsula now need to assess any developments which are proposed for the peninsula in terms of how these may add or detract from the universal significance of the landscape. Establishing clear policies and protocols for excavating and preserving the human remains of those who died in the battles of Gallipoli, developing methods for conflict archaeology with a focus on World War One, and protecting the underwater archaeological heritage that belongs to Gallipoli are only a few of the essential steps in this process.

Valuing the significance of the heritage of pre- and post-WWI, and looking at the landscape through new lenses such as gender, may also bring promising discoveries and new interpretations of Gallipoli. The events of World War One that took place on the Gallipoli peninsula between 1914 and 1916 were of tremendous importance, and the loss of lives on all sides was tragic. How the stories of these events are told and retold in the future will continue to change to reflect the different biases, beliefs, and divergent concerns of the storytellers. That is the fundamental nature of heritage. While it may be naive to expect a shared heritage on the Gallipoli peninsula, one that is less divided and contested could be possibile in the future. The Vimy Declaration for the Conservation of Battlefield Terrain warns us that, "By reason of its nature and history, battle-field terrain is conducive to reflection, contemplation and learning and all too

Figure 6.12 The evacuation from the peninsula. The inscription at the top of the photo reads: "From the Abdurrahman hill, the officers of the Seventh Armed Forces watch the enemy trenches on the morning of the Allied evacuation from the peninsula."

Credit: AWM A05297

attractive to ideological expropriation." Ultimately, if the cultural landscape of the Gallipoli peninsula has taught us any lessons, it is that the winds of politics change quickly, and we should be careful not to destroy the very land that so many cherish (Figure 6.12).

> Bend down and lend your ear, for this silent mound is the place where the heart of a nation sighs.[116]

Notes

1 In Turkish the mission statement is "Milli birlik ve beraberliğimizin harcı olan Çanakkale Savaşları'nın yaşandığı Gelibolu Tarihi Alanı'nın manevi, tarihi, kültürel ve doğal değerlerini, geleceğe olan sorumluluk bilinci ve evrensel düşünceyle gazi köyleri ile birlikte koruyup geliştirerek alan'ı bir açık hava müzesine dönüştürmek." It can be found on the website for ÇATAB at http://catab.kulturturizm.gov.tr/TR,173376/misyonumuz--vizyonumuz.html. ÇATAB came into existence on 28 June 2014 with Law 6546 announced in the Official Gazette 29044.

2 In Turkish the vision statement is "Birliğimizin ve beraberliğimizin sembolü olan Çanakkale Ruhu, dün olduğu gibi gelecekte de var olacaktır" http://catab.kulturturizm.gov.tr/TR,173376/misyonumuz--vizyonumuz.html.

3 For the 2016–2020 Strategic Plan (in Turkish) see http://catab.kulturturizm.gov.tr/Eklenti/50049,2016–2020-stratejik-planpdf.pdf?0.

4 ÇATAB, 2017 Yılı 6 Aylık Raporu (Mid-year Report for 2017), 7. http://catab.kulturturizm.gov.tr/Eklenti/53920,2017-yili-6-aylik-faaliyet-raporu-catabpdf.pdf?0

5 There is a vast amount of literature about the role of the Gallipoli battles in the identity formation of the nations of Australia and New Zealand and the subsequent nationalist pilgrimage to Gallipoli that developed as a result. Among the authors who have written extensively on these topics and whose works are in the bibliography of this book are K. Inglis, D. Lloyd, J. Macleod, B. Scates, P. Unwalla, B. West, B. Ziino.

6 M. Okur, "The Meaning of Gallipoli in Turkish National Identity," in *The Gallipoli Campaign: The Turkish Perspective*, ed. M. Gürcan and R. Johnson (London and New York: Routledge, 2016), 173.

7 E. Yeneroğlu and A. Aykaç, "Battlefield Tourism at Gallipoli: The Revival of Collective Memory, the Construction of National Identity and the Making of a Long-Distance Tourism Network," *AlmaTourism Special Issue* 7, 5 (2016), 69. In "Contemporary Turkish Perceptions of the Gallipoli Campaign," K. Şakul cites that one-third of the 1,914 students in a 2011 survey of middle school Turkish students was not aware of the fact that the Ottomans had been defeated in World War One and that 18 March 1915 was the date of the victory of the Ottomans and the end of World War One. For this survey see Fatma Çoban, "Orta Öğretim Öğrencilerinin Çanakkale Savaşları ve Öğretiminde Biliş ve Algılama Düzeyleri" (unpublished MA thesis, (Çanakkale 18 Mart University, 2011), 94.

8 Turkish Ministry of Forestry and Water Affairs, *Çanakkale Destanı Tanıtım Merkezi* (Ankara, 2012), 5.

9 The "Geography of our Martyrs" project, mentioned in the previous chapter, has been criticized by many scholars of the Çanakkale Wars for historical inaccuracies, and it is to the credit of the Ministry of Forestry and Water Affairs that in the General

Evaluation section of a catalog they published in 2005, the Minister, Osman Pepe and General Director of the Gallipoli Historic National Park, Mustafa Yalınkılıç, mention that the strategy for the symbolic cemeteries in general may be flawed. "We know the names of our martyrs but, unfortunately, we do not actually know in which grave they have been lying. What is more bitter is that we have not marked even the actual martyrs' cemeteries inside and nearby the HNP . . . We have been thinking that the approach of symbolic martyrs' cemeteries is wrong, deviates the historical realities and that offering interpreted and misleading information means both disrespect to our martyrs and makes adverse effect(s) on the meaning and identity of battlegrounds as well. For this reason, it is possible to find out and improve our actual martyrs' cemeteries by using 1916 Şevket Pasha Map and then followed by the work of 'Our Martyrs Geography' to mark the actual martyrs' names along the actual trench lines where they had actually had a fight." *Gallipoli Peninsula Historical National Park* (Ankara, 2005), 47.

10 Yılmaz Öney, "Çanakkale'ye Ruhani Akın," *Milliyet*, 29 July 2004; M. Okur, "The Meaning of Gallipoli," 176.

11 "Gelibolu'da alkol yasağı," *Hürriyet* 19 June 2014 (www.hurriyet.com.tr/gelibolu-da-alkol-yasagi-26640823).

12 Brad West, "Enchanting Pasts: The Role of International Civil Religious Pilgrimage in Reimagining National Collective Memory," *Sociological Theory*, vol. 26, no. 3 (2008), 258–70.

13 Tonie Holt and Valmai Holt, *Gallipoli: Major and Mrs. Holt's Battlefield Guide* (South Yorkshire: Leo Cooper, 2000), 8.

14 *ANZAC DAY: Gallipoli 25 April 2011 Commemorative Services* (Canberra: Dept of Veterans' Affairs, 2011), 1–68.

15 Ibid., 15.

16 M. İredesel, "Çanakkale Savaşları ve Savaş Alanları Tarihi Milli Parkı," Broşür 2 (Gelibolu: Gelibolu Türizm Yayınları, n.d.).

17 Salim Dağ, "Çanakkale Geçilmez Şehitlik Gezi Rehberi" (The Straits shall not be Breached Martyrs' Tour Guide) (Gelibolu, n.p. 2010).

18 Lucienne Thys-Şenocak, "Divided Spaces, Contested Pasts: The Gallipoli Peninsula Historical National Park," in *Public Nature: Scenery, History and Park Design*, ed. Ethan Carr, Shaun Eyring, and Richard G. Wilson (Charlottesville: University of Virginia Press, 2013), 136.

19 R. Yüzük, "Şehitlikler Gezi Rehberi," *Büyük Zafer Gazetesi* (Gelibolu, n.d.). The line of hadith quoted here from a monument on the peninsula is "Şehitler Allah'tan Şunu İstediler: Ya rabbi . . . Bizi Dünyaya Tekrar Gönder ve Senin Uğrunda Bir Kere Daha Şehit Olalım" which is likely a modified version of the following hadith: "Ümmetime ağır gelmeyecek olsaydı, hiçbir seriyyeden geri kalmaz, hepsine katılırdım. Allah yolunda şehit olmak, sonra diriltilip tekrar şehit olmak yine diriltilip tekrar şehit olmak isterdim." (Buhârî, Îman, 26; Müslim, İmâre, 103, 107).

20 For information about OPET's projects on the Gallipoli peninsula see: www.opet.com.tr/en/Respect-to-History; for the Akbaş commemorative project see www.opet.com.tr/opet-tarihe-saygi-projesi-kapsaminda-insa-edilen-akbas-sehitligine-resmi-acilis. OPET states on their website that since 2012 they have helped the peninsula by painting the houses in its villages, organizing the village squares, building three commemorative monuments, paving 85,000 square meters of the peninsula with stone, planting 2850 trees, 18,000 flowers, and 6000 square meters of grass.

21 Sıdıka Eser and Tekin Bilgin, *Gelibolu tarihi alanı* (Ankara: Kültür ve Turizm Bakanlığı, 2016).

22 ÇATAB, *Gelibolu Tarihi Alanı Ziyaret Adabı*. Ecebat, n.d. (http://catab.kulturturizm.gov.tr/Eklenti/48391,ziyaret-adabi-guncellenmis-halipdf.pdf?0&_tag1=DFF0745BE053A6992D067AC89E211194256324AA).

23 D. Lloyd, *Battlefield Tourism: Pilgrimage and the Commemoration of the Great War in Britain, Australia and Canada 1919–1939* (Oxford: Berg, 1998), 40.

24 The audio guide provided for the eleven-chamber tour of the Legend of Gallipoli simulation center is available in seven languages other than Turkish, including English, German, French, Russian, Japanese, Italian, and Arabic. Other detailed information about the center is available at http://catab.kulturturizm.gov.tr/TR,121090/canakkale-destani-tanitim-merkezi.html.

25 The report indicates no revenues apart from ticket sales, hence this is the source for percentage of costs met. See ÇATAB, *2017 Yılı 6 Aylık Faaliyet Raporu*, 97.

26 ÇATAB, *2017 Yılı 6 Aylık Faaliyet Raporu*, 98. See also http://catab.kulturturizm.gov.tr/Eklenti/57114,2017-yili-faaliyet-raporu-indirmek-icin-tiklayinizpdf.pdf?0 for the entire 2017 activities report.

27 The different categories for ticket prices and entry fee exemptions are available at: http://catab.kulturturizm.gov.tr/Eklenti/41348,tarihi-alanda-yer-alan-muze-tanitim-ve-simulasyon-merke-.pdf?0.

28 "Gelibolu Yarımadası'nda hedef 20 milyon turist!" *TourMag*, 4 March 2017; see www.tourmag.com.tr/gelibolu-yarimadasi-20-milyon-turist-hedefliyor/.

29 Yeneroğlu and Aykaç, 74–8.

30 ÇATAB Law No. 6546 passed on 19/6/2014; www.tbmm.gov.tr/kanunlar/k6546.html.

31 Ibid.

32 A. Yılmaz, "Architectural memorialisation of war: Ars Memoriae and the landscape of Gallipoli battles", unpublished Ph.D. diss. (Izmir Institute of Technology, 2008); see pages 180–215 for a detailed discussion of the different projects submitted for this competition and the Long Term Development Project that was proposed in 1999 by R. Bademli and his team at Middle East Technical University for the peninsula after the competition.

33 Ayhan Balamir, "Jüri Üyeleriyle Görüşme," *Mimarlık* 283 (1998), 32–3.

34 Raci Bademli, "Gelibolu Yarımadası Barış Parkı Projesi", *Mimarlık* 283 (1998): 25–6.

35 A. Yılmaz, "Architectural Memorialisation of War: Ars Memoriae and the Landscape of Gallipoli Battles" unpublished Ph.D. dissertation, (Izmir Institute of Technology, 2008).

36 Yılmaz, 192–3.

37 Ibid. 197.

38 M. Gleen Marcutt, Tony Watkins, Ignasi de Solà-Morales Rubió, Robert Riley, Ahmet Gülgönen, Haluk Alatan and Doğan Kuban, "Evaluation of Prize Winning Project: First Prize," Presentation Plate of First Prized Project of the Peace Park Competition, Archive of Eceabat Directorate.

39 Ibid.

40 "Conkbayırı Mehmetçik Park Anıtı," *Mimarlık* 11, no. 85 (1970): 35–7.

41 UDGP, 85, 88; Yılmaz, 147–59 for further analysis of the Conkbayırı monuments.

42 Apart from Erginbaş, the other jurors were Turgut Cansever, Günseli Aru, Ercüment Kalmık and Levent Aksu; additional jury members were Yaşar Marulyalı and Zühtü Müridoğlu.

43 "Conkbayırı Mehmetçik Park Anıtı," 35–37.

44 Ibid.

45 Yılmaz, 156.

46 Both jury members, Levent Aksu and Ercüment Kalmık, note the positive "museum-like" qualities of the Conkbayırı monuments. See "Conkbayırı Mehmetçik Park Anıtı," 36.

47 This is evident from the eroded footpaths around the monuments.

48 R. Bademli and K. Sarı, *Gelibolu Yarımadası Tarihi Milli Parkı (Barış Parkı) Uzun Devreli Gelişme Planı Çalışmaları Tesbit ve Değerlendirme Raporu*, vol. 1 (Ankara, 2002), 78–9.

49 The parking lot and entrance area to this museum and visitor center were expanded after the architectural competition.

50 In Turkish this was announced as the "Çanakkale Gelibolu Yarımadası Tarihi Milli Parkı Kabatepe Sembolik Şehitliği" and its results were published in a brief announcement entitled "Kabatepe Sembolik Şehitliği Yarışması Sonuçlandı," *Mimarlık* 1984, no. 5, 35. See also, Fulya Aksu, "Mimarlık Yarışmaları: Açık Hava Müzesi olarak Gelibolu." www.arkitera.com (24 February 2014).

51 R. Bademli and K. Sarı, *Gelibolu Yarımadası Tarihi Milli Parkı (Barış Parkı) Uzun Devreli Gelişme Planı Çalışmaları Tesbit ve Değerlendirme Raporu*, vol. 1 (Ankara, 2002), 79.

52 www.milliparklar.gov.tr/kitap/55/55.pdf; *Simulation and Information Center for the Legend of Gallipoli* (Ankara, 2012), 38.

53 This was the "Kabatepe Sembolik Şehitliği Ulusal Proje Yarışması" (Kabatepe National Symbolic Martyrs' Project Competition), the winners of which were Metin Hepgüler and İlhan Şahin.

54 There was also no archaeologist on the jury, yet this point was not a concern for either the Ministry or the Chamber of Architects (TMMOB Mimarlar Odası): "Gelibolu Yarımadası Kabatepe Tanıtım Merkezi Fikir Projesi Yarışması Süreci Durdurulmalıdır," TMMOB *Almanak* 09 (15 January 2009), 47; In the book prepared for the opening of the center by the Ministry of Forestry and Water Affairs and the General Directorate of Nature Conservation and National Parks, the writers note that "the competition was open to the individuals and institutions working, doing research or studying in the field of (the) movie industry, 3D movies, animation, picture, arts, and sculpture. The members of the TMMOB were also given (the) chance to enter the competition." *Simulation Center and Information Center for the Legend of Gallipoli* (Ankara, 2012), 13.

55 http://vitruvius.arch.metu.edu.tr/gallipoli/gallipoli_turkish.html; see R. Bademli, ed., *Gelibolu Yarımadası Tarihi Milli Parkı* vol. 2 (2001); Esaslar ve Tutumlar/Fundamentals and Attitudes, ODTÜ, Gelibolu Yarımadası Barış Parkı Planlama ve Danışma Bürosu, AGUDÖS Proje No 99.02.02.03, Ankara.

56 For images of the simulation center and the project drawings see "Gelibolu Yarımadası Tarihi Milli Parkı, Kabatepe Tanıtım Merkezi," at www.arkiv.com.tr/proje/gelibolu-yarimadasi-tarihi-milli-parki-kabatepe-tanitim-merkezi/899lerle; see also the PEB architectural firm's website: www.pebmimarlik.com/. The animation for the center was developed by the Turkish audiovisual company Benart which was responsible for the content production, 3D models, film projections, AV system design, integration and installation. http://benart.net/TR/News/Gallery.aspx?NewsID=236&GalleryType=1.

57 See www.milliparklar.gov.tr/kitap/55/55.pdf and www3.milliparklar.gov.tr/kitap/55/mobile/index.html for the *Simulation and Information Center for the Legend of Gallipoli* (Ankara, 2012), 27.

58 www.akicraatlar.com/uccasnqsceicyxt.html/icraat/canakkale-destani-tanitim-merkezi/22995.

59 www.arkiv.com.tr/proje/gelibolu-yarimadasi-tarihi-milli-parki-kabatepe-tanitim-merkezi/899lerle.

60 Ibid.

61 Ministry of Forestry and Water Affairs/General Directorate of Nature Conservation and National Parks, "The Themes shown in the Center," *Simulation and Information Center for the Legend of Gallipoli* (Ankara, 2012) www.milliparklar.gov.tr/kitap/55/55.pdf. The expression 'funda' or 'fondo bismillah' is used along with 'vira bismillah' as a religious invocation when casting an anchor at sea. See www.terimleri.com/denizcilik/fundo.html.

62 Ministry of Forestry and Water Affairs/General Directorate of Nature Conservation and National Parks, "Nusret Minelayer Ship and Cevat Pasha's Dream," *Simulation and Information Center for the Legend of Gallipoli* (Ankara, 2012), 89. www.milliparklar.gov.tr/kitap/55/55.pdf. (The note at the bottom of this page attributes this information to Prof. Azmi Süslü, Magazine of Atatürk Research Center, 306 but there is no further information for this source.)

63 Ibid.

64 Ibid.

65 Ministry of Forestry and Water Affairs/General Directorate of Nature Conservation and National Parks, "The 8th Chamber: The Battles of Dardanelles from the Dome of (the) Sky," *Simulation and Information Center for the Legend of Gallipoli* (Ankara, 2012), 22. www.milliparklar.gov.tr/kitap/55/55.pdf.

66 E. Erickson, *Ordered to Die: A History of the Ottoman Army in the First World War* (Westport CT: Greenwood Publishing, 2001).

67 *Simulation and Information Center for the Legend of Gallipoli* (Ankara, 2012), 88.

68 Vedica Kant, "Çanakkale's Children: The Politics of Remembering the Gallipoli Campaign in Contemporary Turkey,"*Remembering the First World War*, ed. Bart Ziino (London and New York: Routledge, 2015), 159.

69 Lucienne Thys-Şenocak and Elifgül Doğan, "Archaeology, Museums and Tourism on the Gallipoli Peninsula: Issues of Human Remains, Ordnance, and Local Decision-Making/Gelibolu Yarımadası'nda Arkeoloji, Müzeler ve Turizm: İnsan Kalıntıları, Mühimmat ve Yerel Karar Alma," in *Heritage in Context 2, Miras 4,* (Istanbul: German Archaeological Institute, 2018), 289–313..

70 The term "women worthies" and the phrase "add-women-and-stir" describe an earlier feminist methodology which tries to bring women into a historical narrative perceived as androcentric. It has been criticized, particularly by second and third wave feminists and historians of gender, for not disrupting the analytical frameworks which have been used traditionally to research and interpret gender roles and which are considered to be heavily biased towards elite white males. See Merry Wiesner-Hanks, *Gender in History* (Oxford: Blackwell, 2001), 1.

71 Burcu Çevik-Compiègne, "From Gallipoli (1915) to the War of Independence (1919–1922): Modernisation of Turkish Womanhood," *Cosmopolitan Civil Societies Journal*, 6 no. 3 (2014), 102–15. For an analysis of some of the World War One exhibitions held during the centenary in Australia see Megan Hicks, "Tracing the

Wounded: A Reflection on Six First World War Exhibitions in Australian Museums," *Health and History*, vol. 17, no. 2, Special Issue: World War I (2015), 88–97. The Department of Veteran Affairs in Australia has developed a wealth of teaching materials about the role of Australian Women in World War One which can be found at https://anzacportal.dva.gov.au/file/1013/download?token=KblCtzsY.

72 D. Lloyd, *Battlefield Tourism: Pilgrimage and the Commemoration of the Great War in Britain, Australia and Canada, 1919–1939* (Oxford and New York: Berg, 1998), 45–6.

73 Adam Jones, "imagi(ni)ng gender and conflict," *Feminist Review* no. 101 Conflict (2012), 140.

74 Yılmaz, 130.

75 For the story of Helle see the introduction to this book and Apollodorus, *The Library, with an English Translation by Sir James George Frazer, F.B.A., F.R.S. in 2 Volumes* (Cambridge, MA: Harvard University Press; London: William Heinemann, 1921).

76 L. Thys-Şenocak, *Ottoman Women Builders: The Architectural Patronage of Hadice Turhan Sultan* (Aldershot: Ashgate, 2006), 107–185.

77 Ibid., 112–13.

78 Ibid., 112–13.

79 Ibid., 179. For the oral history project conducted from 1999–2002 at Seddülbahir and Kumkale on the Asian shore of the Dardanelles, see L. Thys-Şenocak and I.C. Cenker, "Moving Beyond the Walls: The Oral History of the Ottoman Fortress Villages of Seddülbahir and Kumkale," in *Oral History and Public Memory*, ed. P. Hamilton and L. Shopes (Philadelphia: Temple University Press, 2008), 65–86.

80 Şafak Horzum, "Reflections on the Gallipoli Campaign in Turkish Literature," *The Gallipoli Campaign: The Turkish Perspective*, ed. M. Gürcan and R. Johnson (London and New York: Routledge, 2016), 123–4.

81 Hüseyin Nihal Atsız, *Çanakkale'ye Yürüyüş, İstanbul*, 1933. This was recently reprinted for the 2015 centenary commemorations by İrfan Publishers in İstanbul. https://turkcetarih.com/canakkaleye-yuruyus/.

82 "UN envoys record WWI poems for 100th anniversary," Associated Press, (4 August 2014). http://newsok.com/un-envoys-record-wwi-poems-for-100th-anniversary/article/feed/718779. The poem was read again at the UN to commemorate an opening of an exhibition for the centenary of World War One.

83 For the original sound recording of Ecevit's poem see https://soundcloud.com/ww1-exhibit-at-the-un/gallipoli-a-post-war-epic-by-bulent-ecevit-read-by-the-turkish-pr-to-the-un-yasar-cevik-1.

84 Jenny Macleod, *Gallipoli* (Oxford: Oxford University Press, 2015), 175.

85 Tony Abbott, "2015 Dawn Service, Gallipoli." Full transcript available at https://pmtranscripts.pmc.gov.au/release/transcript-24397.

86 The translation of the poem and the video of the commercial is available at www.hurriyetdailynews.com/turkish-president-erdogans-gallipoli-prayer-stirs-debate-81350 "Turkish President Erdoğan's Gallipoli 'prayer' stirs debate," *Hürriyet Daily News* (21 April 2015).

87 For the full official text in English of the speech delivered at the Peace Summit by President Erdoğan on 23 April 2015, see "All Armenian Allegations are Baseless," www.tccb.gov.tr/en/news/542/31986/all-armenian-allegations-are-baseless.html.

88 Jenny Macleod, "The Gallipoli Centenary," *War Memory and Commemoration*, ed. B. West (New York and London: Routledge, 2017), 98.

89 "President Serzh Sargsyan responds to Turkish President's invitation letter," Press release of the official web page for the President of Armenia at www.president.am/en/press-release/item/2015/01/16/President-Serzh-Sargsyan-to-the-President-of-Turkey/.

90 This is clearly evidenced in the 18 March 2018 ÇATAB website whose new theme "Birileri Var" (which roughly translates into "Someone is there" and implies that there are many who are defending the homeland) equates the martyrs of the Çanakkale Wars, the Turkish War of Independence, and the Battle of Kut, in Iraq (another battle in which the Ottomans defeated the British on 29 April 1916) with the martyrs of the 15 July attempted coup. See ÇATAB's website for the 18 March 2018 theme (http://18martetkinlikleri.com/Goals).

91 For the entire recording and transcript of this speech delivered on 22 Sept 2016, see www.tccb.gov.tr/en/15-july/1823/52392/july-15-coup-attempt-and-peoples-victory.html.

92 For the official English translation of President Erdoğan's speech made on 18 March 2017, see "Presidential System will pave the way for new Çanakkale Victories," www.tccb.gov.tr/en/news/542/73398/turk-milletinin-yeniden-dirilisi-kacinilmazdir.html.

93 "July 15 Is Our New Çanakkale," www.tccb.gov.tr/en/news/542/78902/15-temmuz-bizim-yeni-bir-canakkalemizdir-sakaryamizdir.html.

94 Ibid.

95 This timeframe is mentioned in both the 2016–2020 Strategic Plan and the "2017 Yılı 6 Aylık Faaliyet Raporu," the (2017 Six Month Activities Report) produced by ÇATAB, available at http://catab.kulturturizm.gov.tr/Eklenti/53920,2017-yili-6-aylik-faaliyet-raporu-catabpdf.pdf?0.

96 Speech delivered on 18 March 2016 by President Erdoğan in commemoration of the Turkish Veterans day, "101 years ago, our nation wrote an epic of resurrection in Çanakkale." www.tccb.gov.tr/en/news/542/41182/canakkale-a-struggle-for-independence-and-the-future.html.

97 *Çanakkale Savaşları Gelibolu Tarihi Alanı 1/25,000 ölçekli tarihi alan planı* produced by Modül Planlama, Ankara. See www.modulplanlama.com.tr/Calismalar/Index/2.

98 "2017 Yılı 6 Aylık Faaliyet Raporu," 95.

99 Sıdıka Eser and Tekin Bilgin, *Gelibolu tarihi alanı* (Ankara: Kültür ve Turizm Bakanlığı, 2016).

100 2016–2020 ÇATAB strategic report, 88. See also *Stratejik Planlama ve Yönetim Çalıştayı*: 29 Mart 2015, no. 2 (Çanakkale: ÇATAB, 2015) for the concluding text of the speech given 29 March 2015 at the ÇATAB Strategic Planning and Management Workshop by the first director of ÇATAB, Mehmet Gürkan.

101 The Ağadere project also included the construction of a commemorative wall 5 meters high and 140 meters long, and a path, the "never-ending road" made of steel and wood which connects the complex's interior structures to the sea in a way that the architect describes as "respectful to nature." For the description of the Ağadere project by the architect see "1. Ödül, Gelibolu Yarımadası Tarihi Milli Parkı Ağadere Hastanesi ve Şehitliği Fikir Projesi Yarışması" http://www.arkiv.com.tr/proje/1-odul-gelibolu-yarimadasi-tarihi-milli-parki-agadere-hastanesi-ve-sehitligi-fikir-projesi-yarisması/741. For the description of the Akbaş project see OPET's website www.opet.com.tr/opet-tarihe-saygi-projesi-kapsaminda-insa-edilen-akbas-sehitligine-resmi-acilis.

102 "Alçıtepe Camisi İbadete Açıldı" 18 March 2018 at www.haberler.com/alcitepe-camisi-ibadete-acildi-10670029-haberi/.

103 Michael Polonsky, John Hall, Julian Vieceli, Lüfti Atay, Ali Akdemir and Mehmet Marangoz, "Using strategic philanthropy to improve heritage tourist sites on the Gallipoli peninsula, Turkey: community perceptions of changing quality of life and of the sponsoring organization," *Journal of Sustainable Tourism* 21:3 (2012), 376–95, doi: 10.1080/09669582.2012.699061. For the legal investigation into OPET's actions see the *Turizm Gazetesi*, 28 August 2006 and "OPET'in "Tarihe Saygı Projesi'ne Soruşturma" at www.tayproject.org/haberarsiv20068.html.

104 Azra Bayraktar and Çağla Cömert, "Effect of Corporate Cultural Responsibility on a Company's Financial Performance and Brand Image: An Example of a Turkish Oil Company Opet A.Ş," in *Corporate Social Responsibility for Valorization of Cultural Organizations*, ed. Maria del Pilar et al. (Hershey, PA: ICI Global, 2018), 216.

105 Mithat Atabay and Turan Takaoğlu, "On the Fate of Ottoman Cultural Properties during the Gallipoli Campaign," *Çanakkale Araştırmaları Türk Yıllığı* 23 (2017), 253, 247–64.

106 2016–2020 ÇATAB strategic report, 111–12.

107 There is also a charge for these guides ranging from 150 tl to 250 tl depending on the size of the tour bus and number of occupants in the bus. See http://catab.kulturturizm.gov.tr/TR,196827/2018-yilinda-alan-kilavuzu-kartina-uygulanacak-bandrol-.html.

108 For the announcement of the "Gallipoli Historic Region New Martyrs Concept Design Project Competition" (Gelibolu Tarihi Şehitlik Alanı Yeni Tasarımları Fikir Projesi Yarışması) http://sehitlikleryarismasi.com/hakkinda/. The 15 sites which are to be commemorated in the competition are: Havantepe, Aslantepe, Küçük Anafarta, Kanlıköprü Deresi, İbrikçe1, İbrickçe 2, Abanos Deresi, İsmailoğlu Deresi, Naimsırtı, Süngübayırı, Albayraksırtı, Kılıçdere, Kiremitdere, Eroğlusırtı.

109 See Nilüfer Karakoç, "1. Ödül, Gelibolu Tarihi Alanı Yeni Şehitlik Tasarımları Fikir Projesi Yarışması," *Arkitera* 25 Aralık/December, 2017 see www.arkitera.com/proje/8618/1-odul-gelibolu-tarihi-alani-yeni-sehitlik-tasarimlari-fikir-projesi-yarismasi-.

110 F. Matero and J. Peters, "Survey Methodology for the Preservation of Historic Burial Grounds and Cemeteries," *APT Bulletin*, vol. 34, no. 2/3 (2003), 38; see also Fred B. Kniffen, "Necrogeography in the United States," *Geographical Review* 57 (1967): 426–7; Paul Groth, "Frameworks for Cultural Landscape Study," *in Understanding Ordinary Landscapes*, ed. Paul Groth and Todd W. Bressi (New Haven: Yale University Press, 1997) 3–21.

111 See www.nycgovparks.org/parks/the-high-line for information on the High Line and www.nps.gov/vive/index.htm for the Vietnam Veterans War Memorial in Washington, DC.

112 David Lloyd, *Battlefield Tourism: Pilgrimage and the Commemoration of the Great War in Britain, Australia and Canada* (Oxford: Berg, 1998), 100.

113 J. Price, "Cultural landscape of sacrifice, the problem of sacred ground of the Great War 1914–1918", Forum UNESCO 10th International Seminar, "Cultural Landscapes in the 21st Century", April 2005

114 For the Burra Charter see: http://australia.icomos.org/wp-content/uploads/The-Burra-Charter-2013-Adopted-31.10.2013.pdf.

115 For the Spirit of Place see: http://whc.unesco.org/uploads/activities/documents/activity-646-2.pdf.

116 "Bir Yolcuya" [For a Traveler], written by Necmettin Halil Onan in 1960.

Bibliography

Abbott, Tony. "2015 Dawn Service, Gallipoli." Full transcript available at https://pmtranscripts.pmc.gov.au/release/transcript-24397.

Akcan, Esra and Sibel Bozdoğan. *Turkey: Modern Architectures in History*. London: Reaktion Books, 2012.

Akın, Yiğit. "Savaş, Milliyetçilik, Sivil Toplum: Çivi Anıtların İstanbul'daki Örneği Hatıra-ı Celadet Topu." *Toplumsal Tarihi*, 243 (March 2014): 62–7.

Alford, Jonathan, ed. *Greece and Turkey: Adversity in Alliance*. Aldershot, UK: Gower, 1984.

Anadolu Gazetesi. "Mezarı 102 Yıl Sonra Gün Yüzüne Çıktı." 2 February 2017.

ANZAC DAY: Gallipoli 25 April 2011 Commemorative Services. Canberra: Dept of Veterans' Affairs, 2011.

Apollodorus. *The Library, with an English Translation by Sir James George Frazer, F.B.A., F.R.S. in 2 Volumes*. Cambridge, MA: Harvard University Press, 1921.

Aslan, Carolyn and Göksel Sazcı. "Across the Hellespont: Maydos (ancient Madytos), Troy and the North-eastern Aegean in the Late Eighth to Early Sixth Century BC." *Annual of the British School at Athens*, 111 (2016): 121–62.

Associated Press. "Erdoğan Disputes Treaty Setting Greek-Turkish Borders, NATO Patrols." *The National Herald*, 6 June, 2016. www.thenationalherald.com/137044/erdogan-disputes-treaty-setting-greek-turkish-borders/.

Atabay, Mithat. "Mübadelede Çanakkale." In *Çanakkale I: Savaşı ve Tarihi*, ed. İ. Yumuşak, 60511. İstanbul: Büyükşehir Belediyesi Kültür ve Turizm Daire Başkanlığı, 2006.

Atabay, Mithat. "Cumhuriyet Döneminde Çanakkale'ye Göçler." In *Çanakkale Tarihi*, ed. Mustafa Demir, VI: 3315–32. İstanbul: Değişim Yayınları, 2008.

Atabay, Mithat. *Atatürk ve Çanakkale*. Edirne: Paradigma Akademi Yayınları, 2015.

Atabay, Mithat. *Gelibolu Yarımadası'nda Yazılan Tarih Yıkılan Anıtlar*. İstanbul: Paradigma Akademi, 2017.

Atabay, Mithat and Turan Takaoğlu. "On the Fate of Ottoman Cultural Properties during the Gallipoli Campaign." *Çanakkale Araştırmaları Türk Yıllığı*, 23 (2017): 247–64.

Atsız, Hüseyin Nihal. *Çanakkale'ye Yürüyüş*. İstanbul: İrfan, 2015.

Ayliffe, Rosie, Marc Dublin and John Gawthrop, *Turkey: The Rough Guide*. London: Harrap, Columbus, 1991.

Bademli, Raci. "Gelibolu Yarımadası Barış Parkı Projesi Uluslararası Fikir ve Tasarım Yarışmasıve Sonrası."*Mimarlık* 283(1998):25–26.http://dergi.mo.org.tr/dergiler/4/516/9229.pdf.

Bademli, Raci and K. Sarı. *Gelibolu Yarımadası Tarihi Milli Parkı (Barış Parkı) Uzun Devreli Gelişme Planı Çalışmaları Tesbit ve Değerlendirme Raporu*, vol. I. Ankara, 2002.

Balamir, Ayhan. "Jüri Üyeleriyle Görüşme." *Mimarlık*, 283 (1998): 30–4.

Baram, Uzi and Lynda Carroll, ed. *A Historical Archaeology of the Ottoman Empire: Breaking New Ground*. Berlin and New York: Springer Science & Business Media, 2006.

Bayhan, Bahar. "Çanakkale Savaşları Gelibolu Tarihi Alan Başkanlığı Kuruluyor." *Arkitera*, 20 June 2014. www.arkitera.com/haber/21751/canakkale-savaslari-gelibolu-tarihi-alan-baskanligi-kuruluyor.

Baykut, Sibel. "The Re-contextualization of the Battle of Gallipoli through Commemorations." *Intercultural Understanding*, 6 (2016): 7–15.

Bayraktar, Azra and Çağla Cömert. "Effect of Corporate Cultural Responsibility on a Company's Financial Performance and Brand Image: An Example of a Turkish Oil Company Opet A.Ş." *Corporate Social Responsibility for Valorization of Cultural Organizations,* ed. Maria del Pilar et al. 186–224. Hershey, PA: ICI Global, 2018.

Baytop, Asuman. *Türkiye'de Botanik Tarihi Araştırmaları*. Ankara: Türkiye Bilimsel ve Teknolojik Araştırma Kurumu, 2005.

Bean, Charles Edwin Woodrow. *The Official History of Australia in the War of 1914–1918: The Australian Imperial Force During the Allied Offensive, 1918*. Vols. 1 and 2. Brisbane: University of Queensland Press, 1983.

Bean, Charles Edwin Woodrow. *Gallipoli Mission*. Canberra: Australian War Memorial, 1948.

Bean, Charles Edwin Woodrow, ed. *The Anzac Book: Written and Collected in Gallipoli by the Men of Anzac*. London: Cassell, 1916.

Becker, Annette. "From Death to Memory: the National Ossuaries in France after the Great War." *History and Memory*, 5, 2 (1993): 32–49.

Becker, Jean Jacques. *The Great War and the French People*. New York: St. Martin's Press, 1986.

Birkett-Rees, Jessie. "Geospatial Science and the Archaeology of the First World War: Context for Conflict." *Geospatial Science Research 2* (GSR_2) at RMIT University (2012). http://ceur-ws.org/Vol-1328/GSR2_BirkettRees.pdf.

Birkett-Rees, Jessie. "Capturing the Battlefield: Mapping and Air Photography at Gallipoli." In *Anzac Battlefield: A Gallipoli Landscape of War and Memory*, ed. Antonio Sagona, Mithat Atabay, C.J. Mackie, Richard Reid, and Ian McGibbon. 49–50. Cambridge: Cambridge University Press, 2016.

Boardman, John. *The Greeks Overseas*. London: Penguin, 1964.

Borg, Alan. *War Memorials: From Antiquity to the Present*. London: Leo Cooper, 1991.

Bozdoğan, Sibel. *Modernism and Nation Building: Turkish Architectural Culture in the Early Republic*. Seattle, WA: University of Washington Press, 2001.

Bozdoğan, Sibel and Akcan, Esra. *Turkey: Modern Architectures in History*. London: Reaktion Books, 2012.

The Burra Charter: The Australia ICOMOS Charter for Places of Cultural Significance, 2013.

Bury, John Bagnell, Stanley Arthur Cook, and Frank E. Adcock. *The Cambridge Ancient History*. 2nd ed. Cambridge: Cambridge University Press, 1933.

Büyüksaraç, Aydın, Cahit Çağlar Yalçıner, Yunus Levent Ekinci, Alper Demirci, and Mehmet Ali Yücel. "Geophysical Investigations at Ağadere Cemetery, Gallipoli Peninsula, NW Turkey." *Australian Journal of Forensic Sciences*, 46, 1 (2014): 111–23.

Callister, Sandy. *The Face of War: New Zealand's Great War Photography*. Auckland: Auckland University Press, 2013.

Cameron, David, and Denise Donlon. "A Preliminary Archaeological Survey of the ANZAC Gallipoli Battlefields of 1915." *Australasian Historical Archaeology*, 23 (2005): 131–8.

Canada Veterans Affairs. "Vimy Declaration for the Conservation of Battlefield Terrain." (2014). www.veterans.gc.ca/eng/remembrance/memorials/overseas/first-world-war/france/vimy/declaration#Preamble.

Çanakkale Destanı Tanıtım Merkezi. Ankara: Turkish Ministry of Forestry and Water Affairs, 2012.

"Çanakkale Zafer ve Meçhul Asker Abidesi Yarışması (I. Ödül)." *Mimarlık*, Ankara, 3 (1944): 52–74.

"Çanakkale Şehitliklerine Ziyaret." *Cumhuriyet*, 13 Ağustos/13 August 1930.

ÇATAB. *Stratejik Planlama ve Yönetim Çalıştayı*. 29 March 2015, No. 2. Çanakkale: ÇATAB, 2015.

ÇATAB. *2017 Yılı 6 Aylık Raporu (Mid-year Report for 2017)*. ÇATAB. *Gelibolu Tarihi Alanı Ziyaret Adabı*. Eceabat, n.d.

Çevik-Compiègne, Burcu. "From Gallipoli (1915) to the War of Independence (1919–1922): Modernisation of Turkish Womanhood." *Cosmopolitan Civil Societies Journal*, 6, 3 (2014): 102–15.

Chasseaud, Peter, and Peter Doyle. *Grasping Gallipoli: Terrain, Maps and Failure at the Dardanelles, 1915*. Staplehurst: Spellmount, 2005.

Chevalier, Nicole. *La recherche archéologique française au Moyen-Orient, 1842–1947*. Paris: Éditions Recherche sur les Civilizations (ERC), 2002. www.bcin.ca/Interface/openbcin.cgi?submit=submit&Chinkey=388630.

Clarke, Hyde. "The Right of Property in Trees on the Land of Another, as an Ancient Institution." *The Journal of the Anthropological Institute of Great Britain and Ireland*, 19 (1890): 199–211.

Clayton, Anthony. *Paths of Glory: The French Army, 1914–1918*. London: Cassell Military, 2003.

Cleary, Michelle. "Anzac Gallipoli Archaeological Database: Appendix." In *Anzac Battlefield: A Gallipoli Landscape of War and Memory*. ed. Antonio Sagona, et al. Cambridge: Cambridge University Press, 2016.

Çoban, Fatma. "Orta Öğretim Öğrencilerinin Çanakkale Savaşları ve Öğretiminde İlişkin Biliş ve Algılama Düzeyleri." Unpublished Master's Thesis, 18 March, Çanakkale University, 2011.

"Conkbayırı Mehmetçik Park Anıtı." *Mimarlık*, 11, 85 (1970): 34–44.

Connell, R.W. and James W. Messerschmidt. "Hegemonic Masculinity: Rethinking the Concept." *Gender and Society*, 19, 6 (2005): 120–32.

Cook, John Manuel. *The Troad: An Archaeological and Topographical Study*. Oxford: Oxford University Press, 1973.

Cosgrove, Denis E. "Introduction to Social Formation and Symbolic Landscape." In *Landscape Theory*, ed. R. DeLue and J. Elkins. New York and London: Routledge, 2008.

Cosgrove, Denis E. *Social Formation and Symbolic Landscape*. 2nd ed. Madison, WI: University of Wisconsin Press, 1998.

Courby, Fernand, Joseph Chamonard, and Edouard Dhorme. "Corps Expéditionnaire d'Orient: Fouilles Archéologiques sur l'Emplacement de la Nécropole d'Éléonte de Thrace." *Bulletin de Correspondance Hellénique*, 39, 1 (1915): 135–240.

Couto, Dejanirah. "Çanakkale Boğazı'ndaki Osmanlı Hisarları/Les fortifications ottomanes des Dardanelles", Üç Denizin Arasında: Osmanlı ve Fransız Boğaz Haritaları/Entre Trois Mers: Cartographie ottomane et française des Dardanelles et du Bosphore, ed. J. Perouse and F. Günergün, 104–17. Izmir: Arkas Sanat Mekezi, 2016.

Cox, Thomas R. "Americans and Their Forests: Romanticism, Progress, and Science in the Late Nineteenth Century." *Journal of Forest History*, 29, 4 (1985): 156–68.

T.C. Cumhurbaşkanlığı. "Çanakkale'den Getirilen Toprak ve Bayrak Cumhurbaşkanı Erdoğan'a Sunuldu." 15 October 2014. www.tccb.gov.tr/haberler/410/1430/canak kaleden-getirilen-toprak-ve-bayrak-cumhurbaskani-erdogana-sunuldu.html.

Cunningham, Matthew. "'Familiarising the Foreign': New Zealand Soldiers' Observations on Landscape during the Gallipoli Campaign." *New Zealand Journal of History*, 45, 2 (2011): 209–24.

Dağ, Salim. *Çanakkale Geçilmez Şehitlik Rehberi* ("The Straits Shall Not be Breached" Martyrs' Tour Guide). n.p. Gelibolu, 2010.

Demangel, Robert. *Le Tumulus Dit de Protésilas*. Paris: E. de Boccard, 1926.

De Nardi, Sarah. *The Poetics of Conflict Experience: Materiality and Embodiment in Second World War Italy*. London and New York: Routledge, 2017.

Department of Veterans Affairs and Board of Studies, Teaching and Educational Standards, New South Wales. "Nurses at Gallipoli." *Gallipoli and the Anzacs* (April 2015). www.gallipoli.gov.au/nurses-at-gallipoli/nurses-experience.php.

Doyle, Peter, and Matthew R. Bennett. "Military Geography: The Influence of Terrain in the Outcome of the Gallipoli Campaign, 1915." *The Geographical Journal*, 165, 1 (1999): 12–36.

Dursun, Selçuk. "Forest and the State: History of Forestry and Forest Administration in the Ottoman Empire." Ph.D. diss., Sabancı University, 2007. http://risc01.sabanciuniv.edu/record=b1164366%20(Table%20of%20Contents).

Dutton, David "'Docile Supernumerary': A French Perspective on Gallipoli," *Gallipoli: Making History*, ed. J. McLeod, 86–97. London and New York: Frank Cass, 2004.

Eldem, Edhem. "From Blissful Indifference to Anguished Concern: Ottoman Perceptions of Antiquities, 1799–1869." In *Scramble for the Past: A Story of Archaeology in the Ottoman Empire, 1753–1914*. ed. Zainab Bahrani, Zeynep Çelik, and Edhem Eldem, 281–329. İstanbul: SALT, 2011.

Eldem, Edhem. "State, Nation and Death." In *Death in Istanbul: Death and its Rituals in Ottoman-Islamic Culture*. İstanbul: Ottoman Bank Archive and Research Centre, 2005.

Erickson, Edward J. *Ordered to Die: A History of the Ottoman Army in the First World War*. Westport, CT: Greenwood, 2001.

Eser, Sıdıka and Tekin Bilgin. *Gelibolu Tarihi Alanı*. Ankara: Kültür ve Turizm Bakanlığı, 2016.

Fasih, Mehmed. *Gallipoli 1915 Bloody Ridge (Lone Pine) Diary of 5th Imperial Ottoman Army Gallipoli 1915*. İstanbul: Denizler Kitabevi, 1997.

Forster, Edward S. "Trees and Plants in Homer." *The Classical Review*, 50, 3 (1936): 97–104.

Fuchs, Ron. "Sites of Memory in the Holy Land: The Design of the British War Cemeteries in Mandate Palestine." *Journal of Historical Geography*, 30 (2004), 643–64.

Gegner, Martin and Bart Ziino, ed. *The Heritage of War*. New York: Routledge, 2012.

Gelibolu Peninsula Historical National Park, Turkish Ministry of Environment and Forestry General Directorate of Nature Protection and National Parks, Ankara: 2005.

"Gelibolu'da alkol yasağı." *Hürriyet*. June 19, 2014. www.hurriyet.com.tr/gelibolu-da-alkol-yasagi-26640823.

"Gelibolu Yarımadası'nda hedef 20 milyon turist!" *TourMag*, 4 March 2017.

"Gelibolu Yarımadası Kabatepe Tanıtım Merkezi Fikir Projesi Yarışması Süreci Durdurulmalıdır." TMMOB *Almanak*, 9, 15 January 2009.

Genelkurmay Askeri Tarih ve Stratejik Etüt Başkanlığı. *Birinci Dünya Harbinde Türk Harbi V*. Ankara: Genelkurmay Başkanlığı Yayınları, 1998.

Gilchrist, R. *Gender and Archaeology: Contesting the Past*. London: Routledge, 1999.

Gill, David W.J. "Excavating Under Gunfire: Archaeologists in the Aegean During the First World War." *Public Archaeology*, 10, 4 (2011): 187–99.

Goode, James. "Archaeology and Diplomacy in the Republic of Turkey, 1919–39." *Turkish-American Relations: Past, Present and Future*, ed. Mustafa Aydın and Çağrı Erhan, 49–65. London: Routledge, 2004.

Goode, James. *Negotiating for the Past: Archaeology, Nationalism, and Diplomacy in the Middle East, 1919–1941*. Austin, TX: University of Texas Press, 2007.

Gooding, Janda. *Gallipoli Revisited: In the Footsteps of Charles Bean and the Australian Historical Mission*. Melbourne: Hardie Grant, 2009.

Greatrex, Geoffrey. "Procopius and Agathias on the Defenses of the Thracian Chersonese." *Constantinople and Its Hinterland: Papers from the Twenty-Seventh Spring Symposium of Byzantine Studies, Oxford April 1993*, ed. Gilbert Dagron and Cyril Mango, 109–24. Aldershot, UK: Variorum, 1995.

Greenhalgh, Michael. "French Military Reconnaissance in the Ottoman Empire during the Eighteenth and Nineteenth Centuries as a Source for Our Knowledge of Ancient Monuments." *The Journal of Military History*, 66 (2002): 359–88.

Grelot, Guillaume-Joseph. *A Late Voyage to Constantinople*. London: J. Playford, 1683.

Groth, Paul. "Frameworks for Cultural Landscape Study." *Understanding Ordinary Landscapes*, ed. Paul Groth and Todd W. Bressi. 3–21. New Haven, CT: Yale University Press, 1997.

Güçlü, Kamuran and Faris Karahan, "A Review: The History of Conservation Programs and Development of the National Parks Concept in Turkey." *Biodiversity and Conservation*, 13 (2004): 1373–90.

Güran, Tevfik. *Osmanlı Devleti'nin İlk İstatistik Yıllığı 1897/The First Statistical Yearbook of the Ottoman Empire 1897*, vol. 5. Ankara: Devlet İstatistik Enstitüsü, 1997.

Gürcan, Metin. "Talking to Hearts and Minds: Influencing Strategies in the Gallipoli Campaign." In *The Gallipoli Campaign: The Turkish Perspective*, ed. Metin Gürcan and Robert L. Johnson, 125–62. London: Routledge, 2016.

Hanna, Martha. *The Mobilization of Intellect: French Scholars and Writers during the Great War*. Cambridge, MA: Harvard University Press, 1996.

Hardgrave, John. *At Suvla Bay: Being the Notes and Sketches of Scenes, Characters and Adventures of the Dardanelles Campaign*. New York: Houghton Mifflin Company, 1917.

Harrington, Alexandra R. "Uncertain Rest: The Nebulous Legal Status of Overseas Military Cemeteries." 11 April 2012. Available at SSRN: https://ssrn.com/abstract=2038389 or http://dx.doi.org/10.2139/ssrn.2038389.

Harrington, Simon. "Human Remains." In *Anzac Battlefield: A Gallipoli Landscape of War and Memory*. ed Antonio Sagona, Mithat Atabay, C.J. Mackie, Richard Reid, and Ian McGibbon. Cambridge: Cambridge University Press, 2016: 93–4.

Hatip, Salih M. "Çanakkale Boğazı (Bahr-i Sefid Boğazı) Savunma Sistemi 1770–1918." Unpublished Ph.D. dissertation, Hacettepe University, Ankara, 2013.

Heyningen, Eleanor van. "Helles: The French at Gallipoli." In *The Joint Imperial War Museum/Australian War Memorial Battlefield Study*. 1–11, London: Imperial War Museum, 2000.

Hicks, Megan. "Tracing the Wounded: A Reflection on Six First World War Exhibitions in Australian Museums." *Health and History, Special Issue: World War I*, 17, 2 (2015): 88–97.

Hill, Aaron. *A Full and Just account of the Present State of the Ottoman Empire in all its branches: with the government and policy, religion, customs, and way of living of the Turk, in general.* London: John Mayo, 1709. http://name.umdl.umich.edu/ 004869666.0001.000.

Hoffenberg, Peter. "Landscape, Memory and the Australian War Experience, 1915–18." *Journal of Contemporary History*, 36, 1 (2001): 111–31.

Holt, Tonie, and Valmai Holt. *Major & Mrs Holt's Battlefield Guide to Gallipoli.* South Yorkshire: Leo Cooper, 2000.

Horzum, Şafak. "Reflections on the Gallipoli Campaign in Turkish Literature." In *The Gallipoli Campaign: The Turkish Perspective*, ed by Metin Gürcan and Robert L. Johnson, 205–16. London: Routledge, 2016.

Howard, Peter. *Heritage: Management, Interpretation, Identity.* London: Continuum, 2003.

Howard, Peter. *Introduction of Landscapes.* Burlington, UK: Ashgate, 2011.

Hughes, D.M. "The French Army at Gallipoli." *The RUSI Journal*, 150, 3 (2005): 64–7.

Hunt, John Dixon. *Greater Perfections: The Practice of Garden Theory.* Philadelphia, PA: University of Pennsylvania Press, 2000.

İğdemir, Uluğ. *Atatürk ve Anzaklar.* Ankara: Türk Tarih Kurumu, 1978.

İnan, Afet. *Atatürk Hakkında Hatıralar ve Belgeler.* İstanbul: Türkiye İş Bankası Kültür Yayınları, 1984.

Inglis, Kenneth Stanley. "War Memorials: Ten Questions for Historians." *Guerres Mondiales et Conflits Contemporains,* No. 167, Les Monuments Aux Morts de la Premiere Guerre Mondiale, (July 1992): 5–21.

Inglis, Kenneth Stanley. "Entombing Unknown Soldiers: From London and Paris to Baghdad." *History and Memory*, 5, 2 (1993): 7–31.

Inglis, Kenneth Stanley, and Jan Brazier. *Sacred Places: War Memorials in the Australian Landscape.* Melbourne: The Miegunyah Press, 2008.

İredesel, M. *Çanakkale Savaşları ve Savaş Alanları Tarihi Milli Parkı, Broşür 2.* Gelibolu: Gelibolu Turizm Yayınları, n.d.

Isaac, Benjamin H. *The Greek Settlements in Thrace until the Macedonian Conquest* (Studies of the Dutch Archaeological and Historical Society, vol. 10). Leiden: Brill, 1986.

Isom-Verhaaren, Christine. "Shifting Identities: Foreign State Servants in France and the Ottoman Empire." *Journal of Early Modern History*, 8, 1 (2004): 109–34.

Jones, Adam. "Imagi(ni)ng gender and conflict." *Feminist Review* 101 *Conflict* (2012): 132–41.

Jong, Adriaan de and Mette Skougaard. "Early Open-Air Museums: Traditions of Museums about Traditions." *Museum International*, 44, 3 (1992): 151–7.

"July 15 Is Our New Çanakkale." www.tccb.gov.tr/en/news/542/78902/15-temmuz-bizim-yeni-bir-canakkalemizdir-sakaryamizdir.html.

Kant, Vedica. "Çanakkale's Children: The Politics of Remembering the Gallipoli Campaign in Contemporary Turkey." In *Remembering the First World War*, ed. Bart Ziino, 146–64. London: Routledge, 2015.

Karataş, Murat. *Haritalarla Çanakkale Savaşları: Gelibolu Yarımadası Kuzey Bölgesi Kara Muharebeleri/Maps of the Çanakkale Wars: Battles in the North Region of the Gallipoli Peninsula.* Ankara: Nobel, 2007.

Karpat, Kemal, *Osmanlı Nüfusu (1830–1914) Demografik ve Sosyal Özellikleri.* Trans. B. Tırnakçı. İstanbul: Tarih Vakfı Yurt Yayınları, 2003.

Kemaloğlu, Azer. "Anniversary and Production of Fiction: Gallipoli." *War Memory and Commemoration*, ed. Brad West, 189–99. London and New York: Routledge, 2017.

Kenyon, Frederic. *War Graves: How the Cemeteries Abroad will Be Designed*. London: His Majesty's Stationery Office, 1918.

Kerin, Lindy. "Indigenous Soldiers Head to Gallipoli for Calling Ceremony." ABC News, 29 July 2015. www.abc.net.au/news/2015–07–29/indigenous-soldiers-travel-to-gallipoli/6656978.

Kersel, Morag. "The Changing Legal Landscape for Middle Eastern Archaeology in the Colonial Era, 1800–1930." *Pioneers to the Past: American Archaeology in the Middle East, 1919–1920*, ed. G. Emberling, 85–90. Chicago, IL: Oriental Institute of the University of Chicago, 2010.

King, Alex. "Remembering and Forgetting in the Public Memorials of the Great War." In *The Art of Forgetting*, ed. Adrian Forty and Susanne Küchler. London: Berg, 2001.

Kniffen, Fred. "Necrogeography in the United States." *Geographical Review*, 57 (1967): 426–7.

Knopp, A. Bernard and Wendy Ashmore. "Archaeological Landscapes: Constructed, Conceptualized, Ideational." In *Archaeologies of Landscape: Contemporary Perspectives*, ed. A. Knopp and W. Ashmore, 1–30. Oxford: Blackwell, 1999.

Koca, Yasemin. "Dardanelles, Hellespont, Çanakkale: Çanakkale Haritaları." *Çanakkale Araştırmaları Türk Yıllığı*, 15, 22 (2017): 109–34.

Kolay, Selçuk and Savaş Karatay. "Echoes from the Deep: Wrecks of the Dardanelles Campaign." In *Underwater Cultural Heritage from World War I: Proceedings of the Scientific Conference on the Occasion of the Centenary of World War I*. Bruges: UNESCO Publishing, 2015.

Kolay, S., O. Taktak, S. Karataş, and M. Atabay. *Echoes from the Deep: Wrecks of the Dardanelles Campaign*. İstanbul: Vehbi Koç Foundation/Ayhan Şahenk Foundation, 2013.

Köprülü, Bülent. "Türk Hukukunda Orman Rejimi." *İstanbul Üniversitesi Hukuk Fakültesi Mecmuası*, 14, 3–4 (1948): 700–54.

Körpe, Reyhan. "Archaeological Surveys at the Battlefields of Gallipoli, Çanakkale Savaş Alanlarında Yapılan Arkeolojik Araştırmalar." In *100. Yılında Çanakkale Savaşları Uluslararası Kongresi/ International Congress on Battles of Gallipoli on the Centenary*, ed. A. Akengin and O. Neçare, 773–94. Ankara: Atatürk Araştırma Merkezi Yayınları, 2018.

Kreiser, Klaus. "Public Monuments in Turkey and Egypt, 1840–1916." *Muqarnas*, 14 (1997): 103–17.

Kreiser, Klaus. "Public Monuments in Kemalist and Post-Kemalist Turkey." *Türklük Bilgisi Araştırmaları/Journal of Turkish Studies*, 26, 11 (2002): 43–60.

Kritovoulous, *The History of Mehmed the Conqueror*. trans. C.T. Riggs. Princeton, NJ: Princeton University Press, 1954.

Kulu, Mustafa. "1530'da Çanakkale Yöresinde Yerleşme ve Nüfus." *Çanakkale Araştırmaları Türk Yıllığı*, 2, 2 (2004): 197–239.

Kutluk, Halil. "Türkiye'de Yabancı Ormancılar." In *Türk Ormancılığı Yüzüncü Tedris Yılına Girerken 1857–1957*. Ankara: Türkiye Ormancılar Cemiyeti, 1957.

Laffin, John. "Gallipoli the Largest British Memorial in the World." *Royal British Legion Journal*, March 1974.

Lambert, George. *Autobiography of George Lambert*. Mitchell Library, Sydney, ML MSS A1811.

Laqueur, Thomas W. "Memory and Naming in the Great War." In *Commemorations: The Politics of National Identity*, ed. John R. Gillis, 150–67. Princeton, NJ: Princeton University Press, 1994.

Laqueur, Thomas W. *Making Sex: Body and Gender from the Greeks to Freud*. Cambridge, MA: Harvard University Press, 1990.

Leaf, Walter. *Troy: A Study in Homeric Geography*. London: Macmillan, 1912.

Leaf, Walter. "The Military Geography of the Troad." *The Geographical Journal*, 47, 6 (1916): 401–16.

Lecane, Philip. *Beneath a Turkish Sky: The Royal Dublin Fusiliers and the Assault of Gallipoli*. Gloucestershire, UK: History Press, 2015.

Lees, Justin. "Half-Naked Female Sniper Stalked Anzacs at Gallipoli. Just a Battlefield Rumour?" *NewsComAu*, 18 June 2015. www.news.com.au/national/anzac-day/halfnaked-camouflaged-women-snipers-stalked-the-anzacs-at-gallipoli-more-than-a-myth/news-story/3b48ba23d11c737fa89bc8e80af46fa9.

Lefebvre, Henri. *The Production of Space*. Trans. Donald Nicholson-Smith. Cambridge, MA: Basil Blackwell, 1991.

Lloyd, David William. *Battlefield Tourism: Pilgrimage and the Commemoration of the Great War in Britain, Australia and Canada, 1919–1939*. Oxford: Berg, 1998.

Longworth, Philip. *The Unending Vigil: A History of the Commonwealth War Graves Commission, 1917–1984*. South Yorkshire, UK: Leo Cooper, 2003.

Lowenthal, David. *Perkins March: Prophet of Conservation*. Seattle, WA: University of Washington Press, 2015.

Lowenthal, David. "Past Time, Present Place: Landscape and Memory." *Geographical Review*, 65, 1 (1975): 1–36.

Lyndon, Jane. "Young and Free: The Australian Past in a Global Future." In *Cosmopolitan Archaeologies*, ed. Lynn Meskell, 28–47. Durham, NC and London: Duke University Press, 2009.

McCarthy, Justin. *Muslims and Minorities: The Population of Ottoman Anatolia at the End of the Empire*. New York: New York University Press, 1983.

McCarthy, Justin. *Population History of the Middle East and the Balkans*. İstanbul: Isis Press, 2002.

McKenna, Mark and Stuart Ward. " 'It Was Really Moving, Mate': The Gallipoli Pilgrimage and Sentimental Nationalism in Australia." *Australian Historical Studies*, 129 (2007): 141–51.

Mackie, C. J. "Archaeology at Gallipoli in 1915." In *Philathenaios, Studies in Honour of Michael J. Osborne*. ed. Anastasios Tamis, C. Mackie, and S. Byrne. Athens, Greece: Greek Epigraphic Society (Anatypon) (2010): 213–25.

Mackie, C. J., Mithat Atabay, Reyhan Körpe, and Antonio Sagona. "Boundary and Divide: The Antiquity of the Dardanelles." In *Anzac Battlefield: A Gallipoli Landscape of War and Memory,* ed. Antonio Sagona, Mithat Atabay, C. J. Mackie, Richard Reid, and Ian McGibbon. Cambridge: Cambridge University Press, 2016.

Macleod, Jenny. *Reconsidering Gallipoli*. Manchester and New York: Manchester University Press, 2004.

Macleod, Jenny. ed. *Gallipoli: Making History*. London and New York: Taylor and Francis, 2004.

Macleod, Jenny. *Gallipoli*. Oxford: Oxford University Press, 2015.

Macleod, Jenny. "The Gallipoli Centenary an International Perspective." In *War Memory and Commemoration*, ed. Brad West, 89–106. London: Routledge, 2017.

McNeill, John Robert. *The Mountains of the Mediterranean World*. Cambridge: Cambridge University Press, 2002.

McQuilton, John. "Gallipoli as Contested Commemorative Space." In *Gallipoli: Making History*, ed. Jenny Macleod, 150–8. London and New York: Frank Cass, 2004.

Marc, J. "L'archeologie francaise en Thrace: A la mémoire des tirailleurs sénégalais du 54éné régiment colonial," Actes du deuxième symposium international des Etudes Thraciennes. *Thrace ancienne*, Komotini, Grèce (1997): 473–506.

Marquez-Grant, Nicholas and Linda Fibiger. *The Routledge Handbook of Archaeological Human Remains and Legislation: An International Guide to Laws and Practice in the Excavation and Treatment of Archaeological Human Remains*. London: Routledge, 2011.

Marsh, George P. *Man and Nature*, ed. D. Lowenthal. Cambridge, MA: Belnap Press, 1965.

Masefield, John. *Gallipoli*. New York: Macmillan, 1917.

Matero, Frank and Judy Peters. "Survey Methodology for the Preservation of Historic Burial Grounds and Cemeteries." *APT Bulletin*, 34, 2/3 (2003): 37–45.

Matthieu, René-Hubert. "Des Hellénistes en Guerre." *Revue Historique des Armées*, 261 (2010): 1–19.

Mayo, James M. "War Memorials as Political Memory." *Geographical Review*, 78 (1988): 62–75.

Meskell, Lynn. "Cosmopolitan Heritage Ethics." *Cosmopolitan Archaeologies*, ed. Lynn Meskell, 1–27, Durham, NC and London: Duke University Press, 2009.

Midford, Sarah. *Gallipoli, Anzacs and the Great War*. Melbourne: La Trobe University Press, 2017.

Mikhail, Alan. *Nature and Empire in Ottoman Egypt: An Environmental History*. Cambridge: Cambridge University Press, 2011.

Millington, Tasman. *French War Cemetery Cape Helles*. Imperial War Graves Commission. Çanakkale: IWGC, 1947.

Milliyet. n.a. "Abide için yardım 1.5 milyonu aştı." 6 April 1958.

Moorehead, Alan. *Gallipoli 1915*. New York: Ballantine Books, 1956.

Moriarty, Catherine. "Christian Iconography and the First World War Memorials," *Imperial War Museum Review*, 6 (1992): 63–75.

Nelson, S. *Handbook of Gender in Archaeology*. Berkeley, CA: AltaMira, 2006.

Nizamoğlu, Serdar. Paper presented at the Third Strategies and Activities Conference for the Çanakkale Wars and Gallipoli Historical Region Plans and Site Management Plan, Kolin Hotel, Çanakkale, 23 February 2017.

Nora, Pierre. "Between Memory and History: Les Lieux de Mémoire." *Representations*, 26 (1989): 7–24.

Nordenson, Eva. "In the Beginning . . . Skansen." *Museum International*, 44, 3 (1992): 149–50.

O'Keeffe, Tadhg. "Landscape and Memory: Historiography, Theory, Methodology." *Heritage, Memory and the Politics of Identity: New Perspectives on the Cultural Landscape*, ed. Niamh Moore and Yvonne Whelan, 3–18. Aldershot, UK: Ashgate, 2007.

Okur, Mehmet A. "The Meaning of Gallipoli in Turkish National Identity." In *The Gallipoli Campaign: The Turkish Perspective*. ed. M. Gürcan and R. Johnson, 173–80. London: Routledge, 2016.

Olson, Sherry H. "Commerce and Conservation: The Railroad Experience." *Forest History Newsletter*, 9 (1966): 2–15.

Olson, Sherry H. *The Depletion Myth: A History of Railroad Use of Timber*. Cambridge, MA: Harvard University Press, 1971.

Onat, Emin and Orhan Arda. "Anıt-Kabir!" *Arkitekt*, 25, 2 (1955): 55–9. Öney, Yılmaz. "Çanakkale'ye Ruhani Akın." *Milliyet*, 29 July 2004.

Onur, Necmi. *Çanakkale Savaşları ve Şehitler Abidesi*. İstanbul: Gün Matbaası, 1960.

Onur, Necmi. "Şehidler abidesi alaka bekliyor." *Milliyet*, 21 March 1963.

Onur, Oral and Latif Boğman. "Edirne Şehit Anıtları." *Türk Şehitlikleri İmar Vakfi Yayınları*, 2 (2012).

Özbek, Onur. "Hamaylıtarla Reconsidered: A Neolithic Site and Its Environmental Setting in Southern Turkey." *Anatolia Antiqua*, 18 (2010): 1–21.

Özden, Sezgin and Üstüner Birben. "Ottoman Forestry: Socio-Economic Aspect and Its Influence Today." *Ciência Rural*, 42, 3 (2012): 459–66.

Özdoğan, Mehmet. "A Surface Survey for Prehistoric and Early Historic Sites in Northwestern Turkey." *National Geographic Society Research Reports*, 20 (1985): 517–41.

Özdoğan, Mehmet. "Prehistoric Sites in the Gelibolu Peninsula." *Anadolu Araştırmaları*, 10 (1986): 51–67.

Özel, Sibel. "Under the Turkish Blanket Legislation: The Recovery of Cultural Property Removed from Turkey." *International Journal of Legal Information*, 38 (2010). http://scholarship.law.cornell.edu/ijli/vol38/iss2/10.

Özgüner, Nimet Pınar. "Archaeological Entanglements: People, Places, and Politics of Archaeology in Turkey," Ph.D. Unpublished Dissertation, Boston University, MA, 2015.

Özkale, Edip and Mustafa R. Şenler. *Haritacı Mehmet Şevki Paşa ve Türk Haritacılık Tarihi (1919 Yılına Kadar)*. Ankara: Harita Genel Müdürlüğü, 1980.

Özyürek, Esra. *Nostalgia for the Modern: State Secularism and Everyday Politics in Turkey*. London and Durham NC: Duke University Press, 2006.

Pedersen, Peter. *Anzac Treasures: The Gallipoli Collection of the Australian War Memorial*. Sydney: Murdoch Books, 2014.

Pemberton, T. J. *Gallipoli To-Day*. London: E. Benn, 1926.

Perez-Alvaro, Elena. "Human Remains on Underwater Cultural Heritage: Ethics, Values and Law." In *Proceedings of the 2nd Asia-Pacific Regional Conference on Underwater Cultural Heritage*. ed. Hans Van Tilburg, S. Tripati, V. Walker Vadillo, B. Fahy, and J. Kimura. 2014: 39–48.

Picard, Charles and Adolphe-J. Reinach. "Voyage dans la Chersonèse et aux Îles de la Mer de Thrace." *Bulletin de Correspondance Hellénique*, 36, 1 (1912): 275–352.

Piekarz, Mark. "It's Just a Bloody Field! Approaches, Opportunities and Dilemmas of Interpreting English Battlefields." In *Battlefield Tourism: History, Place and Interpretations*, ed. Chris Ryan, 29–48. Oxford: Elsevier, 2007.

Pisani, Donald. "Forests and Conservation, 1865–1890," *The Journal of American History*, 72, 2 (1985): 340–59.

Polonsky, Michael, John Hall, Julian Vieceli, Lütfi Atay, Ali Akdemir and Mehmet Marangoz. "Using strategic philanthropy to improve heritage tourist sites on the Gallipoli peninsula, Turkey: community perceptions of changing quality of life and of the sponsoring organization." *Journal of Sustainable Tourism*, 21, 3 (2012): 376–95.

Piri Reis. *The Book of Bahriye*, ed. Bülent Özükan. İstanbul: Boyut, 2013.

Pottier, Edmond. "Coupe Attique d'Eléonte au Musée du Louvre." *Bulletin de Correspondance Hellénique*, 55 (1931): 430–7.

"President Erdoğan picks up the flag again." *Daily Sabah*, 7 November 2014. www.dailysabah.com/politics/2014/11/07/president-erdogan-picks-up-the-flag-again.

"President Serzh Sargsyan responds to Turkish President's invitation letter." Press release of the official web page for the President of Armenia at www.president.am/en/press-release/item/2015/01/16/President-Serzh-Sargsyan-to-the-President-of-Turkey/.

Price, Jon. "Cultural Landscape of Sacrifice, the Problem of the Sacred Ground of the Great War 1914–1918", Paper Presented at the Forum UNESCO University and Heritage 10th International Seminar, Newcastle, 2005.

Ryan, Chris ed., *Battlefield Tourism: History, Place and Interpretations*. Oxford: Elsevier, 2007.

Reid, Richard. "Recording the Battlefield: First Steps." *Anzac Battlefield: A Gallipoli Landscape of War and Memory*, ed. Antonio Sagona et al., 36–58. Cambridge: Cambridge University Press, 2016.

Reid, Richard and Ian McGibbon. "Forming the Anzac Battlefield." *Anzac Battlefield: A Gallipoli Landscape of War and Memory*, ed. Antonio Sagona, et al., 98–137. Cambridge: Cambridge University Press, 2016.

Sagona, Antonio, Mithat Atabay, C. J. Mackie, Richard Reid, and Ian McGibbon. "An Archaeology of the ANZAC Battlefield." *Humanities Australia*, 6 (2015): 34–46.

Sagona, Antonio, Mithat Atabay, C. J. Mackie, Richard Reid, and Ian McGibbon, ed. *Anzac Battlefield: A Gallipoli Landscape of War and Memory*. Cambridge: Cambridge University Press, 2016.

Sagona, Antonio, and Jessie Birkett-Rees. "Battlefield Archaeology: Gallipoli." *Anzac Battlefield: A Gallipoli Landscape of War and Memory*, ed. Antonio Sagona et al., 83–97. Cambridge: Cambridge University Press, 2016.

Şafak, Horzum. "Reflections on the Gallipoli Campaign in Turkish Literature." In *The Gallipoli Campaign: The Turkish Perspective*, ed. Metin Gürcan and Robert L. Johnson, 125–62. London: Routledge, 2016.

Şakul, Kahraman ed., *Cihan Harbinde Gelibolu ve Mirası/Gallipoli in the Great War and its Heritage Exhibition Catalog*. İstanbul: İstanbul Şehir University, 2016.

Şakul, Kahraman. "Contemporary Perceptions of the Gallipoli Campaign." In *The Gallipoli Campaign: The Turkish Perspective*. ed. M. Gürcan and R. Johnson, 181–204. London and New York: Routledge, 2016.

Salaville, Sévérien. Ambassade de France a Constantinople. "Fouilles du Corps d'Occupation Français de Constantinople Exécutées de 1920 à 1923 par ordre du Général Charpy, avec le Concours Technique de l'École Française d'Athènes sous la Direction de M. Ch. Picard, Publiées par R. Demangel; Th. Macridy-Be; E. Mamboury; Mme M. Massoul; Préface de Ch. Diehl; Fascicule I: Le Tumulus dit de Protésilas, par R. Demangel. Ouvrage publié sous les auspices du ministère de l'Instruction publique et des Beaux-Arts." *Échos d'Orient*, 26 (1927): 494–7.

Sanders, Liman Von. *Five Years in Turkey*. Philadelphia, PA: The Williams and Wilkins Company, 1928.

Saunders, Nicholas. "Excavating Memories: Archaeology and the Great War, 1914–2001." *Antiquity*, 76, 291 (2002): 101–8.

Saunders, Nicholas, ed. *Matters of Conflict: Material Culture, Memory and the First World War*. London: Routledge, 2004.

Saunders, Nicholas. *Killing Time: Archaeology and the First World War*. Stroud, UK: Sutton, 2007.

Saunders, Nicholas, ed. *Beyond the Dead Horizon: Studies in Modern Conflict Archaeology*. Oxford: Oxbow, 2012.

Saunders, Nicholas and Paul Cornish, ed. *Bodies in Conflict: Corporeality, Materiality, and Transformation*. London: Routledge, 2014.

Saunders, Nicholas. "Bodies in Trees: A Matter of Being in Great War Landscapes." In *Bodies in Conflict: Corporeality, Materiality, and Transformation*, ed. Nicholas Saunders and Paul Cornish, 22–38. London: Routledge, 2014.

Saunders, Nicholas and Paul Cornish. "Series Introduction: Material Culture and Modern Conflict." *Conflict Landscapes and Archaeology from Above*, ed. Birger Stichelbaut and David Cowley. London: Routledge, 2016.

Sayar, Zeki. "Anit-Kabir Müsabakası Münasebeti İle." *Arkitekt*, 3–4 (1943): 1–21.

Sazcı, Göksel. "Maydos Kilisetepe Settlement Mound." *City of Legends and Epics Çanakkale*, ed. Özden, 227–30. İstanbul: Yapı Kredi Press, 2012.

Scates, Bruce. *Return to Gallipoli: Walking the Battlefields of the Great War*. Cambridge: Cambridge University Press, 2006.

Schama, Simon. *Landscape and Memory*. Toronto: Random House, 1995.

Schlich, William, Richard Alexander Hess, Johann Karl Gayer and William Rogers Fisher, *Schlich's Manual of Forestry*. London: Bradbury Agnew, 1906.

Schliemann, Heinrich. *Troy and Its Remains: A Narrative of Researches and Discoveries Made on the Site of Ilium, and in the Trojan Plain*. ed. Phillip Smith. London: John Murray, 1875.

Schliemann, Heinrich. *Troja: Results of the Latest Researches and Discoveries on the Site of Homer's Troy and in the Heroic Tumuli and Other Sites, Made in the Year 1882. And a Narrative of a Journey in the Troad in 1881*. London: John Murray, 1884.

Scott, Alexander. "Russell Crowe's Water Diviner Tries to Question History, but Misses the Mark." *Gallipoli, Anzacs and the Great War*. ed. Sarah Midford. Melbourne: La Trobe University Press, 2017.

Scott, Douglas D., Richard A. Fox, Melissa A. Connor, and Dick Harmon. *Archaeological Perspectives on the Battle of the Little Bighorn*. Norman, OK: University of Oklahoma Press, 2013.

Scott, Douglas D. and Andrew P. McFeaters. "The Archaeology of Historic Battlefields: A History and Theoretical Development in Conflict Archaeology." *Journal of Archaeological Research*, 19 (2011): 103–32.

Şehitlerimiz, Vols. I–III. Ankara: Milli Savunma Bakanlığı, 1999.

Sertel, Zekeriya. *Resimli Ay*, 1 Ağustos/August 1341/1925.

Şevki Paşa, Mehmet. *Çanakkale Tahkimat Haritası/Defense Atlas of Çanakkale 1916*. Ankara: Genelkurmay ATASE Başkanlığı Yayınları, 2009.

Sherman, D. J. "Art, Commerce, and the Production of Memory in France after World War I." *Commemorations: The Politics of National Identity*. ed. John R. Gillis, 186–211. Princeton, NJ: Princeton University Press, 1994.

Sherman, D. J. "Bodies and Names: The Emergence of Commemoration in Interwar France." *The American Historical Review*, 103 (1998): 443–66.

Simulation and Information Center for the Legend of Gallipoli. Ankara, 2012.

Slade, Peter. "Gallipoli Thanatourism: The Meaning of ANZAC," *Annals of Tourism Research*, 30, 4 (2003), 779–94.

Smith, Tim. "AE2 The Silent Anzac. Australia's Lost Gallipoli Submarine." *Iron, Steel and Steamship Archaeology, Proceedings of 2nd Australian Seminar: Australian Institute for Maritime Archaeology Special Publication*. ed. M. McCarthy, 139–42. Melbourne, 2006.

Sözen, Metin. "Türklerde Anıt." *Mimarlık*, 7 (July, 1973): 10.

Stanley, Peter. *Quinn's Post: Anzac, Gallipoli*. Sydney: Allen & Unwin, 2005.

Sperling, Jerome W. "Kum Tepe in the Troad: Trial Excavation, 1934." *Hesperia: The Journal of the American School of Classical Studies at Athens*, 45 (1976): 305–64.

Stratejik Planlama ve Yönetim Çalıştayı: 29 Mart 2015, no. 2. Çanakkale: ÇATAB, 2015.

Sudaş, İlknur. "Gelibolu Yarımadası Tarihi Milli Parkı Yeni Şehitlik Tasarımları Fikir Projesi Yarışması Ön Seçim Sonuçları." *Arkitera*, 4 June 2014. www.arkitera.com/haber/21479/gelibolu-yarimadasi-tarihi-milli-parki-yeni-sehitlik-tasarimlari-fikir-projesi-yarismasi-on-secim-sonuclari.

Tanyeri-Erdemir, Tuğba. "Archaeology as a Source of National Pride in the Early Years of the Turkish Republic." *Journal of Field Archaeology*, 31, 4 (2006): 381–93.

Tassy, Louis. *Études sur l'aménagement des Forêts*. Paris: Au Bureau des annales forestières, 1858.

Taylor, Josie. "Danna Vale Suggests Australian Replicate of Gallipoli." *AM Monday*, 18 October 2005. www.abc.net.au/am/content/2005/s1484587.htm.

Thys-Şenocak, Lucienne and Gizem Dörter. "Fortifications in Turkey: The Ottoman Frontier Fortresses of Seddülbahir and Upper Rumeli Kavak." In *Conservation of Cultural Heritage in Turkey*, ed. Z. Ahunbay, D. Mazlum, and Z. Eres/English ed. L. Thys-Şenocak and E. Yıldırım, 356–86. İstanbul: Ege Yayınları/ICOMOS Turkey, 2016.

Thys-Şenocak, Lucienne and Carolyn Aslan. "Narratives of Destruction and Construction: The Complex Cultural Heritage of the Gallipoli Peninsula." In *Archaeology of Destruction*, ed. Lila Rokoczy, 29–47. Cambridge: Cambridge University Press, 2008.

Thys-Şenocak, Lucienne. *Ottoman Women Builders: The Architectural Patronage of Hadice Turhan Sultan*. Aldershot, UK: Ashgate, 2006.

Thys-Şenocak, Lucienne. "Divided Spaces, Contested Pasts: The Gallipoli Peninsula Historical National Park." *Public Nature: Scenery, History and Park Design*, ed. Ethan Carr, Shaun Eyring, and Richard Guy Wilson, 127–40. Charlottesville, VA: University of Virginia Press, 2013.

Thys-Şenocak, Lucienne, Carolyn Aslan and Rahmi N. Çelik, "Gelibolu Yarımadası Tarihsel Savaş Alanı Arkeolojisi Araştırmaları: Seddülbahir Kalesi Projesi (Research of the Historical and Battlefield Archaeology of the Gallipoli Peninsula: The Ottoman Fortress at Seddülbahir)" *TUBA-AR. Turkish Academy of Sciences Journal of Archaeology*, 11 (2008): 107–20.

Thys-Şenocak, Lucienne, Rahmi N. Çelik, and Carolyn Aslan. "Research at the Ottoman Fortress of Seddülbahir, 2005, 2006 Seasons." *International Archaeology, Research and Archaeometry Symposium Proceedings*. Ankara: T.C. Ministry of Culture and Tourism, (2008): 63–76.

Thys-Şenocak, Lucienne and İ. Cerem Cenker. "Moving Beyond the Walls: The Oral History of the Ottoman Fortress Villages of Seddülbahir and Kumkale." In *Oral History and Public Memories*, ed. Paula Hamilton and Linda Shopes, 65–86. Philadelphia, PA: Temple University Press, 2008.

Thys-Şenocak, Lucienne and Elifgül Doğan. "Archaeology, Museums and Tourism on the Gallipoli Peninsula: Issues of Human Remains, Ordnance, and Local Decision-Making/Gelibolu Yarımadası'nda Arkeoloji, Müzeler ve Turizm: İnsan Kalıntıları, Mühimmat ve Yerel Karar Alma." *Heritage in Context*, 2, Miras 4, ed. Katja Piesker et al., 289–313. Istanbul: German Archaeological Institute (DAI), 2018.

Tosyavizade, Rıfat Osman. *Edirne Abideleri Yazıları*. ed R. Duymaz and Y. Topaloğlu, Edirne: Trakya University, 2015.

"Treaty of Lausanne." The World War I Document Archive. 1923. https://wwi.lib.byu.edu/index.php/Treaty_of_Lausanne.

"Treaty of Peace with Turkey, and other Instruments, Signed at Lausanne on July 24, 1923." *Treaty Series*, 16 (1923). http://treaties.fco.gov.uk/docs/pdf/1923/ts0016-1.pdf.

Trout, Steven. "Contested Objects: Material Memories of the Great War." *First World War Studies*, 4 (2013): 261–3.

Tuncel, Bedrettin. *Atatürk ve 30 Ağustos Zaferi'nin İlk Kutlanışı*. Ankara: TTK, 1972.

Türkmen, Z., A. Çalışkan, and F. İlhan. *Çanakkale Savaşı Anıtları ve Şehitlikleri (The Monuments and Martyrs' Graves of the Çanakkale War)*. Ankara: Genelkurmay Basımevi, 2007.

"Turkish President Erdoğan's Gallipoli 'prayer' stirs debate," *Hürriyet Daily News* (21 April 2015).

Türköz, Meltem. "Surname Narratives and the State-Society Boundary: Memories of Turkey's Family Name Law of 1934." *Middle Eastern Studies*, 43, 6 (2007): 893–908.

"UN envoys record WWI poems for 100th anniversary." *Associated Press*. 4 August 2014.

UNESCO *Underwater Cultural Heritage from World War One*. ed. Ulrike Gurin, Arturo Rey da Silva and Lucas Simond, Paris: UNESCO, 2015. www.unesco.org/culture/underwater/world-warI.pdf.

UNESCO. "Underwater Cultural Heritage." Paris: UNESCO, 2001. www.unesco.org/new/en/culture/themes/underwater-cultural-heritage/2001-convention/.

UNESCO. "UNESCO World Heritage Centre; Tentative Lists: Turkey." Paris: UNESCO World Heritage Centre. http://whc.unesco.org/en/tentative lists/state=tr.

UNESCO. "Cultural Landscapes." Paris: UNESCO World Heritage Centre. Accessed 2 June 2017. http://whc.unesco.org/en/culturallandscape/.

Unwalla, Pheroz. "Re-imagining Gallipoli: Imperial Pasts and Foreign Presence in a History of Turkish National Remembrance, 1923–2013" (Ph.D. diss., SOAS, 2014)

Valenti, Catherine. "L'école Française d'Athènes pendant la Grande Guerre: Une institution universitaire au service de l'entente." *Guerres Mondiales et Conflits Contemporains*, 204 (2001): 5–14.

Vassal, Joseph Marguerite Jean. *Uncensored Letters from the Dardanelles*. London: W. Heinemann, 1916.

Vorderstrasse, Tasha. "The Archaeology of the Ottoman Empire and Its Aftermath in the Middle East." *Near Eastern Archaeology*, 77 (2014): 292–8.

Waiblinger, Angelika. "La Ville Grecque d'Éléonte en Chersonèse de Thrace et sa Nécropole." *Comptes Rendus des Séances de l'Académie des Inscriptions et Belles-Lettres*, 122 (1978): 843–57.

Ward, G. Kingsley and Major Edwin Gibson, *Courage Remembered: The Story behind the Construction and Maintenance of the Commonwealth's Military Cemeteries and Memorials of the Wars 1914–1918 and 1939–1945*. London: HMSO, 1988; reprinted 1995.

West, Brad. "Enchanting Pasts: The Role of International Civil Religious Pilgrimage in Reimagining National Collective Memory." *Sociological Theory*, 27, 3 (2008): 258–70.

West, Brad. "Independent Travel and Civil Religious Pilgrimage: Backpackers at the Gallipoli Battlefields." *Down the Road: Exploring Backpacker and Independent Travel*, ed. Brad West, Perth, Australia: API Network/Australia Research Institute/Curtin University of Technology, 2005.

Whelan, Yvonne and Niamh Moore, ed. *Heritage, Memory and the Politics of Identity: New Perspectives on the Cultural Landscape*. New York: Routledge, 2016.

Wiesner-Hanks, Merry. *Gender in History*. Oxford: Blackwell, 2001.

Wilcken, Ulrich. *Alexander the Great*. New York: W.W. Norton, 1997.

Wilson, Christopher. *Beyond Anıtkabir: The Funerary Architecture of Atatürk*. Farnham, UK and Burlington, VT: Ashgate, 2013.

Wilson, Mary. "The Making of Melbourne's Anzac Day." *Australian Journal of Politics & History*, 20, 2 (1974): 197–209.

Winter, Caroline. "First World War Cemeteries: Insights from Visitor Books." *Tourism Geographies*, 13, 3 (2011): 462–79.

Winter, Jay. *Sites of Memory, Sites of Mourning: The Great War in European Cultural History*. Cambridge: Cambridge University Press, 1995.

Wise, Nathan. " 'Dig, Dig, Dig, until You Are Safe': Constructing the Australian Trenches on Gallipoli." *First World War Studies*, 3, 1 (2012): 51–64.

Yeneroğlu, Kutbay E. and A. Aykaç. "Battlefield Tourism at Gallipoli: The Revival of Collective Memory, the Construction of National Identity and the Making of a Long-Distance Tourism Network." *AlmaTourism*, 7, 5 (2016): 61–83.

Yılmaz, Ahenk. "Memorialization on War-Broken Ground: Gallipoli War Cemeteries and Memorials Designed by Sir John James Burnet." *Journal of the Society of Architectural Historians*, 73, 3, (2014): 328–46

Yılmaz, Ahenk. "Architectural Memorialisation of War: Ars Memoriae and the Landscape of Gallipoli Battles." Ph.D. Diss. (Izmir Institute of Technology, 2008).

Yücel, Erdem. "Mimar Muzaffer 1881–1920." *Bizim Anadolu* (4 July 1971).

Yund, Kerim. *Türkiye Orman Umum Müdürleri Albümü*. İstanbul: Hüsnü Tabiat Matbaası, 1959.

Yüzük, R. "Şehitlikler Gezi Rehberi." *Büyük Zafer Gazetesi*. Gelibolu, n.d.

Ziino, Bart. ed. *Remembering the First World War*. London and New York, Routledge, 2015.

Ziino, Bart. "'We are talking about Gallipoli after all': contested narratives, contested ownership and the Gallipoli Peninsula." *The Heritage of War*, ed. Martin Gegner and Bart Zino, 142–59. London and New York: Routledge, 2012.

Ziino, Bart. "Who Owns Gallipoli? Australia's Gallipoli Anxieties, 1915–2005." *Journal of Australian Studies*, 30, 88 (2006): 1–12.

Zürcher, Erik. "The Ottoman Empire and the Armistice of Mudros." In *At the Eleventh Hour: Reflections, Hopes, and Anxieties at the Closing of the Great War, 1918*, ed. Hugh Cecil and Peter H. Liddle, 266–75. London: Leo Cooper, 1998.

Index

Page numbers in *italics* denote an illustration, **bold** a table

For Product Safety Concerns and Information please contact our EU
representative GPSR@taylorandfrancis.com
Taylor & Francis Verlag GmbH, Kaufingerstraße 24, 80331 München, Germany

www.ingramcontent.com/pod-product-compliance
Ingram Content Group UK Ltd.
Pitfield, Milton Keynes, MK11 3LW, UK
UKHW022021180926
475701UK00005B/973
* 9 7 8 0 3 6 7 5 8 7 2 5 3 *